W0232020

BioMethods Vol. 7

Series Editors

Dr. Thomas Meier
Physiologisches Institut
Universität Basel
Vesalgasse 1
4051 Basel
Switzerland

Dr. H.-P. Saluz
Hans-Knöll-Institut
für Naturstofforschung e.V.
Beutenbergstr. 11
D-07745 Jena
Germany

A Laboratory Guide to Biotin-Labeling in Biomolecule Analysis

Edited by
T. Meier
F. Fahrenholz

Birkhäuser Verlag
Basel · Boston · Berlin

Editors

Dr. Thomas Meier
Physiologisches Institut
Universität Basel
Vesalgasse 1
4051 Basel
Switzerland

Dr. F. Fahrenholz
Max-Planck-Institut für Biophysik
Kennedyallee 70
D-60596 Frankfurt
Germany

Library of Congress Cataloging-in-Publication Data

A laboratory guide to biotin-labeling in biomolecule analysis / edited by T. Meier, F. Fahrenholz.
p. cm.
Includes bibliographical references and index.
ISBN-13: 978-3-0348-7351-2 e-ISBN-13: 978-3-0348-7349-9
DOI: 10.1007/978-3-0348-7349-9
1. Biotin-Laboratory manuals. 2. Affinity labeling-Laboratory manuals. 3. Immunoadsorption- Laboratory manuals. 4. Affinity chromatography-Laboratory manuals. I. Meier, T. (Thomas), 1962– . II. Fahrenholz, F. (Falk)
QP772.B55.L33 1996
5 74.19'26-dc20

Deutsche Bibliothek Cataloging-in-Publication Data

A laboratory guide to biotin-labeling in biomolecule analysis /
ed. by T. Meier ; F. Fahrenholz. - Basel ; Boston ; Berlin:
Birkhäuser, 1996
(BioMethods ; Vol. 7)
ISBN-13: 978-3-0348-7351-2
NE: Meier, Thomas [Hrsg.] ; GT

Softcover reprint of the hardcover 1st edition 1996
Printed on acid-free paper produced from chlorine-free pulp. TCF ∞

ISBN-13: 978-3-0348-7351-2

9 8 7 6 5 4 3 2 1

Contents

1 An Introduction to Avidin-Biotin Technology and Options for Biotinylation 1

1.1 **Introduction** 1
The Avidin-Biotin Interaction 1
Avidin 2
Streptavidin 2
Biotin 3
1.2 **Technical Procedures** 3
Determining Biotin and Biotin Incorporation Levels 4
Biotinylation Reagents 5
General Considerations for Biotinylation 7
Amine-Reactive Biotinylation Reagents 8
Water-Soluble NHS Esters of Biotin 9
Water-Insoluble NHS Esters of Biotin 10
General Considerations for the Use of NHS-Ester Biotinylation Reagents 11
Biotinylating IgG with NHS-Ester Biotinylation Reagents 12
Carbohydrate-Directed Biotinylation Reagents 13
Considerations for the Use of Hydrazide Derivatives of Biotin 14
Carboxyl-Reactive Biotinylation Reagents 16
Considerations for the Use of EDC 16
Sulfhydryl-Reactive Biotinylation Reagents 17
Considerations for the Use of Sulfhydryl-Reactive Biotinylation Reagents 18
Biotinylation of Surface Thiols with Biotin-HPDP 19
Biotinylating Reduced IgG with Iodoacetyl-LC-Biotin 20
Biotinylating Reduced IgG with Biotin-BMCC 21
Phenylazide Derivatives of Biotin 22
Considerations for the Use of Photoactivatable Biotin 23
Biotinylation with Photoactivatable Biotin 23
1.3 **Troubleshooting** 24
Avidin-HABA Biotin Assay 24
NHS-Ester Biotinylation Reactions 25
Biotin-Hydrazide Biotinylation Reactions 25
EDC/5-(Biotinamido)pentylamine Reactions 26
Iodoacetyl-LC-Biotin Reactions 26
Biotin-BMCC Reactions 27
Phenylazide Biotinylation Reactions 27

Acknowledgments 27

References 28

2 **Synthesis of Photocleavable Biotinylated Ligands and Application for Affinity Chromatography** ... 31

Summary ... 31

2.1 **Introduction** ... 32
2.2 **Technical Procedures** ... 32
Chemical Synthesis ... 35
2.3 **Results and Discussion** ... 38
Chemical Synthesis ... 38
Binding to anti-CCK antisera and CCK-B receptors ... 40
Binding to streptavidin agarose and photoelution ... 40
Affinity chromatography ... 41
2.4 **Troubleshooting** ... 43

References ... 44

3 **Purification of the Receptor for Pituitary Adenylate Cyclase-Activating Polypeptide (PACAP) using Biotinylated Ligands** ... 45

Summary ... 45

3.1 **Introduction** ... 46
3.2 **Technical Procedures** ... 48
Preparation of a Biotinylated Ligand ... 48
Preparation of Membrane Fraction ... 50
Solubilization of the PACAP Receptor ... 51
Partial Purification of the PACAP Receptor ... 52
Affinity Purification of the PACAP Receptor ... 54
Final Purification of the PACAP Receptor ... 56
Receptor-Binding Assay ... 57
3.3 **Results and Discussion** ... 58
3.4 **Troubleshooting** ... 61
Impurity in PACAP27-Cys-NH_2 ... 61
Low specific activity of the PACAP receptor despite a single band ... 62

Acknowledgments ... 62

References ... 63

4 **Photoreactive Biotinylated Peptide Ligands for Affinity Labeling** ... 65

Summary ... 65

4.1 **Introduction** ... 66
4.2 **Technical Procedures** ... 67
Synthesis of a trifunctional photoactivatable biotinylating reagent ... 67
Synthesis of photoreactive biotinylated peptide hormones ... 67
Site-specific incorporation of biotin and photo labels in separate steps ... 69
Photoaffinity labeling ... 72
4.3 **Results and Discussion** ... 76
Synthesis of photoactivatable insulins with permanent biotin labels ... 76
Applications: Insulin ... 78

Examples for other applications ... 78
4.4 **Troubleshooting** ... 79

References ... 81

5 **Immunoprecipitation of Biotinylated Cell Surface Proteins** ... 83

Summary ... 83

5.1 **Introduction** ... 84
5.2 **Technical Procedures** ... 86
Cell lines ... 86
5.3 **Results and Discussion** ... 90
5.4 **Troubleshooting** ... 93
Sample preparation ... 93
Immunoprecipitation ... 94
Western transfer, membranes, and detection ... 95

Acknowledgments ... 96

References ... 97

6 **Biotinylation and Chemical Cross-Linking of Membrane Associated Molecules** ... 99

Summary ... 99

6.1 **Introduction** ... 100
6.2 **Technical Procedures** ... 101
Chemicals and monoclonal antibodies ... 101
Cell lines and preparation of cell suspensions ... 101
6.3 **Results and Discussion** ... 105
Biotinylation and chemical cross-linking of molecules on the lymphocyte surface ... 105
Permeabilizing cells with lysolecithin to biotinylate and cross-link intracellular molecules ... 107
6.4 **Troubleshooting** ... 112

Acknowledgments ... 113

References ... 113

7 **Preparation of Biotinylated Lectins and Application in Microtiter Plate Assays and Western Blotting** ... 115

Summary ... 115

7.1 **Introduction** ... 116
7.2 **Technical Procedures** ... 117
Biotinylation of lectins ... 118
The microtiter plate assay ... 119
Lectinoblotting ... 121
7.3 **Results and Discussion** ... 122

7.4 Troubleshooting ... 128
Acknowledgements ... 129
References ... 129

8 Biotin-Labeling of Poly(ADP-ribose) in Poly(ADP-ribose)-Protein Interactions ... 131
Summary ... 131
8.1 Introduction ... 132
8.2 Technical Procedures ... 133
Purification of poly(ADP-ribose) polymerase and preparation of poly (ADP-ribose) ... 133
Biotinylation of poly(ADP-ribose) ... 135
Use of biotinylated poly(ADP-ribose) for ligand blotting ... 136
8.3 Results and Discussion ... 137
8.4 Troubleshooting ... 140
Purification and storage of poly(ADP-ribose) polymerase ... 140
Preparation of poly(ADP-ribose) ... 140
Unspecific signals at ligand blotting ... 140
Time considerations ... 140
Acknowledgments ... 141
References ... 141

9 Preparation and Use of Biotinylated Probes for the Detection and Characterisation of Serine Proteinase and Serine Proteinas Inhibitory Proteins ... 143
Summary ... 143
9.1 Introduction ... 144
9.2 Technical Procedures ... 145
Biotinylation procedures ... 146
Compositional and functional analyses conducted on the biotinylated probes . 149
Calculation of IC50 values ... 151
Applications of the biotinylated probes ... 153
9.3 Results and Discussion ... 155
Biotinylated SLPI ... 155
Biotinylated PCTI-1 ... 156
Biotinylated aprotinin ... 156
Biotinylated trypsin ... 158
9.4 Troubleshooting ... 160
General considerations ... 160
Specific considerations ... 162
Acknowledgments ... 164
References ... 162

10 **Avidin/Biotin-Mediated Conjugation of Antibodies to Erythrocytes: An Approach for Immunoerythrocyte Exploration in vivo** 167

Summary 167

10.1 Introduction 168
10.2 Technical Procedures 169
10.3 Results and Discussion 173
Modification of RBC with biotin/streptavidin and attachment of b-IgG to RBC .. 173
Stability of immunoerythrocytes in serum: *in vitro* study 176
Binding of immunoerythrocytes to antigen: *in vitro* study 177
Biodistribution and circulation of serum-stable immunoerythrocytes in rats 178
Conclusion 178
10.4 Troubleshooting 179

Acknowledgments 181

References 182

11 **Biotin *in vitro* Translation: A Nonradioactive Method for the Synthesis of Biotin-Labeled Proteins in a Cell-Free System** 183

Summary 183

11.1 Introduction 184
11.2 Technical Procedures 184
General 184
11.3 Results and Discussion 188
11.4 Troubleshooting 197

Acknowledgments 198

References 198

12 **Nonradioactive Detection of Nucleic Acids with Biotinylated Probes** 201

Summary 201

12.1 Introduction 202
12.2 Technical procedures 202
PCR-labeling of (c)DNA probes 202
Northern blot 204
Southern blot 208
Quantitative analysis by densitometry 208
12.3 Results and discussion 210
12.4 Troubleshooting 212
No or weak signal 212
Background problems 212

References 213

13 Biotin-Labeled Riboprobes to Study RNA-Binding Proteins ... 215

Summary ... 215

13.1 Introduction ... 216
13.2 Technical Procedures ... 217
Preparation of biotinylated riboprobes ... 217
13.3 Results and Discussion ... 220
13.4 Troubleshooting ... 223
Biotinylated riboprobes ... 223
Northwestern assay ... 224

Acknowledgments ... 224

References ... 225

14 Rapid YAC End Sequencing by Alu-Vector PCR and Biotinylated Primers ... 227

Summary ... 227

14.1 Introduction ... 228
14.2 Technical Procedures ... 229
Isolation of YAC DNA ... 229
Alu-vector PCR ... 230
14.3 Results and Discussion ... 234
14.4 Troubleshooting ... 236

Acknowledgments ... 237

References ... 237

Introduction

This book summarizes protocols and applications of recently developed or improved non-radioactive biotin-labeling techniques for proteins, glycoproteins and nucleic acids and will provide valuable help to researchers both in fundamental and in applied sciences.

The first chapter of this volume compares the chemical properties of biotin-labeling compounds currently available and outlines their reaction principles. The following contributions provides a step-by-step protocol on how to prepare and successfully apply biotin-labeled probes for the analysis of complex biochemical and cellular systems. An extended troubleshooting section completes each of the protocols. In most cases these core protocols provide a guideline that encourages modifications according to the researchersí experimental designs.

Combined with sensitive detection, these recently developed experimental procedures are powerful tools for many applications in areas ranging from protein biochemistry to molecular and cellular biology.

Thomas Meier
Denver, USA

Falk Fahrenholz
Frankfurt, Germany

January 1996

Abbreviations

Abbreviations

Ab	antibodies
ACTH	adrenocorticotropic hormone
ADP	adenosine-5'-diphosphate
AEBSF	4-(2-aminoethyl)-benzenesulfonyl fluoride
ε-Ahx	ε-aminohexanoic acid
AIA	biotinylated jacalin
AS	ammonium sulfate
ATP	adenosine-5'-triphosphate
bA	biotinylated aprotinin
bAb	biotinylated antibodies
BANA	4-(Biotin-ε-Ahx-oxymethyl)-3-nitrobenzoyl-Gly-Orn (propionyl)-ε-aminohexanoic acid
BCA	bicinchoninic acid
BCIP	bromochloroindoyl phosphate
BIGCHAP	*N,N*-bis(3-D-gluconamidopropyl) cholamide
BNHS	succinimide ester of biotin
Boc	t-butyloxycarbonyl
bPCTI-1	biotinylated potato chymotrypsin inhibitor-1
bRBC	biotinylated red blood cells
BSA	bovine serum albumin
bSLPI	biotinylated secretory leucocyte proteinase inhibitor
bT	biotinylated trypsin
BxNHS	long-arm biotin ester
CCK	cholecystokinin
CCK-8ds	desulfated cholecystokinin octapeptide
CCK-8s	sulfated cholecystokinin octapeptide
CHAPS	3-[(3-cholamidopropyl)dimethylammonio]propanesulfonic acid
CTP	cytidine-5'-phosphate
dATP	deoxyadenosine-5'-triphosphate
DCC	*N,N'*-dicyclohexyl carbodiimide
dCTP	deoxycytidine-5'-phosphate
DEAE	diethylaminoethyl

DEPC	diethyl pyrocarbonate
dGTP	deoxyguanosine-5'-triphosphate
DMEM	Dulbecco's modified Eagle medium
DMF	*N,N*-dimethylformamide
DMSO	dimethyl sulfoxide
DSP	dithio-bis(succinimidylpropionate)
DTSSP	dithio-bis(sulfo-succinimidylpropionate)
DTT	dithiothreitol (Cleland's reagent)
dTTP	deoxythymidine-5'-triphosphate
dUTP	deoxyuridine-5'-triphosphate
ECL	enhanced chemiluminescence
EDC	1-ethyl-3-(3-dimethylaminopropyl)-carbodiimide hydrochloride
EDTA	etylenediaminetetraacetic acid
EGF	epidermal growth factor
EGF-R	epidermal growth factor receptor
EGTA	etylene glycol-bis(2-aminoethyl ether)*N,N,N',N'*-tetra-acetic acid
EI-MS	electron ionisation mass spectroscopy
ELISA	enzyme linked immunosorbent assay
ELISIA	enzyme linked immunosorbent inhibition assay
EMSA	electrophoretic mobility shift assay
Fab-MS	fast atom bombardement mass spectroscopy
FACS	fluorescence-activated cell sorter
FCS	fetal calf serum
FITC	fluorescein isothiocyanate
Fmoc	9-fluorenylmethoxycarbonyl
GPA	Glycophorin A
GTP	guanosine-5'-triphosphate
GVB	gelatin-veronal buffer
HABA	2-(4'-hydroxyazobenzene)-benzoic acid
Hepes	*N*-(2-hydroxyethyl)piperazine-*N'*-(2-ethanesulfonic acid)
HNTSS	2-hydroxy-5-nitro-α-toluene-sulphonic acid sultone
HOBt	1-hydroxybenzotriazole
HONSu	*N*-hydroxysuccinimide
HPLC	high performance liquid chromatography
HRP	horseradish peroxidase
IgG	immunoglobulin G
MBP	maltose-binding protein
MEA	mercaptoethylamine
MES	*N*-morpholinoethane sulfonic acid
MOPS	3-(*N*-morpholino)propanesulfonic acid
α-MSH	α-melanocyte stimulating hormone
NAD	nicotinamide-adenine dinucleotide
NBT	nitro blue tetrazolium

NHS	*N*-hydroxysuccinimide
NPGB	4-nitrophenyl-4'-guanidinobenzoate
NTP	nucleotide-triphosphate
PACAP	pituitary adenylate cyclase-activating polypeptide
PAGE	polyacryl amide gel electrophoresis
PAMAC	photoaffinity-mediated avidin complexing
PBS	phosphate-buffered saline
PCA	perchloric acid
PCTI-1	potato chymotropsin inhibitor-1
PFGE	pulsed-field gel elctrophoresis
PIPES	piperazine-*N,N'*-bis (2-ethanesulfonic acid)
PMSF	phenyl methyl sulphonyl fluoride
PNA	peanut agglutinin
POD	peroxidase
PVDF	polyvinylidene difluoride
RBC	red blood cells
RES	reticuloendothelial system
R_t	retention time
SA	streptavidin
SAAPPNA	succinyl-Ala-Ala-Pro-Phe-4-nitroanilide
SATA	*S*-acetylthioglycolic acid *N*-hydroxysuccinimide ester
SBTI	soybean trypsin inhibitor
SDS	sodium dodecyl sulfate
SEM	2-(trimethylsilyl)ethoxymethyl
SLPI	secretory leucocyte proteinase inhibitor
SPDP	3-(2-pyridyldithio)propionic acid *N*-hydroxysuccinimide ester
SPIs	serine proteinase inhibitory proteins
SSC	sodium-sodium citrate buffer
SSPE	sodium-sodium phosphate-EDTA buffer
TAE	Tris-acetate-EDTA buffer
TBS	Tris-buffered saline
TBS-T	Tris-buffered saline containing Tween 20
TCR	T cell receptor
TE	Tris-EDTA buffer
TFA	trifluoracetic acid
TLCK	*N*-tosyl-L-lysine chloromethyl ketone
TPCK	*N*-tosyl-phenylalanine chloromethyl ketone
Tris	tris(hydroxymethyl)aminoethane
TTBS	Tris-buffered saline containing Tween 20
UTP	uridine-triphosphate
VVA	vicia villosa agglutinin
WGA	wheat germ agglutinin
ZAPNA	benzyloxycarbonyl-arginine-4-nitroanilide

1 An Introduction to Avidin-Biotin Technology and Options for Biotinylation

M. Dean Savage

1.1 Introduction

Avidin-biotin chemistry represents an enormous toolbox for the biological researcher, owing to the extremely high affinity of biotin for its binding proteins, avidin and streptavidin, as well as the ability to detect the interaction by nonradioactive methods. Early research in the 1920s to 1950s focused primarily on the nutritional implications of biotin and its importance as a coenzyme for carboxylases. Beginning in the early 1970s, however, the avidin-biotin interaction began to be exploited as a research tool, with several techniques being developed in broad areas such as affinity chromatography, blotting, ELISA, hybridization, and others. Specialized techniques, improvements, and new adaptations continue to be developed, limited only by the creativity, vision, and needs of the researcher. It is the intent of this author to provide an overview of avidin-biotin chemistry with particular emphasis on biotinylation reagents that are commonly available. For more detailed reviews, the reader is referred to earlier works (1, 2).

The Avidin-Biotin Interaction

The avidin-biotin interaction is the strongest known noncovalent biological recognition ($K_d = 10^{-15}$ M) between protein and ligand. Bond formation between biotin and avidin is very rapid and, once formed, is unaffected by wide extremes of pH, temperature, organic solvents, or other denaturing agents. The avidin-biotin complex can withstand brief exposure to temperatures up to 132 °C (3, 4), but the temperature stability of the complex is dependent on salt concentration (5, 6). Avidin retains its ability to bind to biotin at room temperature in the presence of common detergents such as SDS, Tween 20, Tween 40, and Triton X-100 (7).

The avidin-biotin complex is not significantly affected by pH values between 2 and 13 or by 8 M guanidine hydrochloride at neutral pH (3). Outside these parameters, the loss of biotin-binding activity is largely due to the denaturation of avidin; the dissociation of avidin into subunits has been reported to be largely reversible (3, 8). A combined treatment with guanidine hydrochloride at pH 1.5, or autoclaving, has been reported to break the avidin-biotin interaction (3, 9). Other researchers have reported that 6 M guanidine·HCl, pH 1.5, only results in 25% dissociation of the avidin-biotin complex (10). The author suggests 8 M guanidine·HCl, pH 1.5, treatment for efficient dissociation. Alternatively, the complex can be boiled in SDS-reducing sample buffer; this is particularly useful for precipitation protocols using immobilized avidin or streptavidin which will be followed by electrophoresis.

Avidin

Avidin is a tetrameric glycoprotein originally isolated from the whites of chicken eggs. This protein has an isoelectric point of approximately 10 (11) and has a molecular weight of about 68 000 (3, 12). The four subunits of avidin are identical; the native tetramer has four biotin-binding sites. The oligosaccharide present on each subunit is a distinguishing characteristic of avidin. The carbohydrate is linked to Asn 17 on the subunit through one of its acetylglucosamine residues (3, 12). Several studies reported that the binding of avidin or streptavidin to biotin is random (3, 13).

Streptavidin

Streptavidin is a 60 000 Da, tetrameric, biotin-binding protein isolated from culture broth of *Streptomyces avidinii* (14, 15). Streptavidin does not possess carbohydrate and has a pH of approximately 5–6. Streptavidin is interchangeable with avidin in many applications, but streptavidin is usually preferred due to its lower nonspecific binding characteristics. Deglycosylated, charge-modified derivatives of avidin, such as Neutravidin (Pierce), have also been developed to decrease nonspecific binding. Streptavidin, like avidin, is highly resistant to denaturation and is even more resistant than avidin to dissociation into subunits by guanidine·HCl (3). Streptavidin has four biotin-binding sites, and cooperative binding of biotin occurs (16). In comparing the sequence of streptavidin with avidin, Gitlin et al. have pointed out that the shared high affinity for biotin by these proteins may be due to the conserved tryptophan-lysine sequences in their primary structures (17). Avidin contains the Trp-Lys sequence in two positions, 70–71 and 110–111, and streptavidin retains this conservation with the Trp-Lys sequences at positions 79–80 and 120–121.

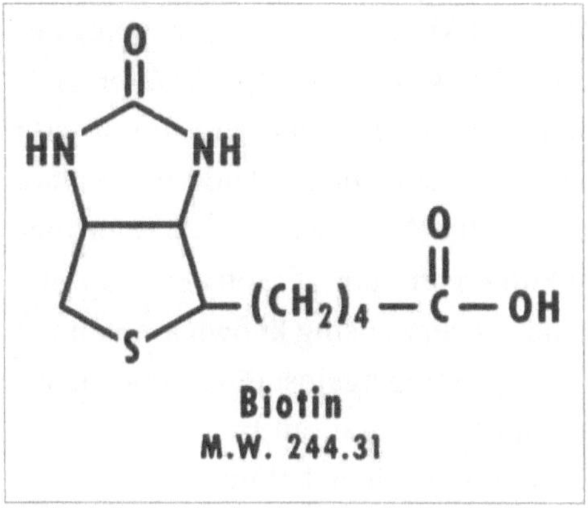

Figure 1.1 Structure of biotin

Biotin

The D-isomer of biotin (Fig. 1) is a naturally occurring vitamin found in every living cell. The tissues with the highest amounts of biotin are the liver, kidney, and pancreas. Yeast and milk are also high in biotin (18). Cancerous tumors have a higher biotin content than does normal tissue (18). Biotin is incompatible with formaldehyde and with oxidizing agents such as chloramine T and nitrous acid (18). Biotin has very few chromophoric properties. At 250 nm, biotin has an absorbance of 0.111/mg/mL (19).

1.2 Technical Procedures

A variety of methods have been used to quantitate biotin and biotinylated compounds. Biotin is difficult to measure by direct spectrophotometric methods, especially in the presence of complex mixtures. However, detection of the carbonyl group of biotin at 220 nm with a high-performance liquid chromatography (HPLC) system using a reversed-phase C18 column, as well as 2 separate ion-exchange columns, has been reported (20). The quenching of tryptophan fluorescence caused by biotin binding to avidin has also been used as the basis for an assay (21). *p*-Dimethylaminocinnamaldehyde can be used to measure biotin in a chemical fashion with a lower limit of 10 μg (22). The probe, 2-anilinonapthalene-6-sulfonate (2,6-ANS), has also been used in a fluorometric assay for both biotin and avidin (23). Perhaps the most commonly used method for assaying the extent of biotin incorporation on a molecule is the HABA method (24). HABA (4'-hydroxyazobenzene-2-carboxylic acid) is a dye that binds with low affinity ($K_d = 6 \times 10^{-6}$ M at pH 4.7) to avidin (3). When HABA is added to a solution of avidin, an orange color with absorption at 500 nm is produced. On addition of biotin or a biotinylated compound, the HABA dye is displaced, causing a decrease in the A_{500} adsorption proportional to the biotin content of the sample. The unknown amount of biotin present in a solution can be determined by preparing a standard curve using known amounts of biotin to displace the HABA bound to avidin and plotting against the change in absorbance at 500 nm. Alternatively, the assay can be carried out by using the extinction for the HABA-avidin complex. This method is given below.

Determining Biotin and Biotin Incorporation Levels

This method is carried out within a single spectrophotometer cuvette to limit the consumption of reagent and sample. Alternatively, the assay can be modified for use with a standard curve.

Materials

1. Avidin-HABA reagent. To prepare, add 8 mg of avidin and 600 μl of 10 mM HABA solution (24.2 mg of HABA in 9.9 ml of H_2O + 0.1 ml of 1 N NaOH) to 19.4 ml of phosphate-buffered saline (PBS). The A_{500} nm of this solution should be approximately 0.9. This solution is stable for about 2 weeks at 4 °C.

If a precipitate forms in the HABA solution, it can be filtered and then used. This will not affect the remainder of the reagent.

2. Biotinylated compound or biotin.

High levels of biotin incorporation on a protein may require pronase digestion, prior to assay, to yield a quantitative value.

3. Spectrophotometer cuvette, 1-ml capacity.

Protocol 1.1

Avidin-HABA Biotin Assay

1. Add 900 µl of the avidin-HABA reagent to a 1-ml cuvette. Record A_{500}.
2. Add 100 µl of the biotinylated sample to the cuvette.
3. Cover the top of the cuvette with Parafilm and invert to mix, and then record A_{500}. (Wait until the A_{500} is stable for at least 15 sec.)

If the A_{500} is less than or equal to 0.3, the sample should be diluted and the assay repeated.

Calculations:

ΔA_{500} = [0.9 (A_{500} of avidin-HABA reagent)] – (A_{500} after addition of biotinylated sample)

µmol of biotin/ml of reaction mixture = (ΔA_{500}/34) µmol/ml

mol of biotin/mol of sample = ((µmol of biotin/ml of reaction)(10))/(µmol/ml concentration of sample)

A multiplier of 10 is used in the above calculation since 90% of the solution is avidin-HABA reagent and 10% is biotinylated sample.

Biotinylation Reagents

Only the intact bicyclic ring of biotin is required for the strong interaction exhibited between avidin and biotin. The carboxyl group on the valeric acid side chain of biotin is not involved in this interaction (3, 25). Consequently, biotin derivatives with reactive properties to a variety of different functional groups can be prepared. These derivatives are commonly called biotinylation reagents and each is prepared through chemistries that modify the valeric acid carboxyl group on native biotin. A variety of biotinylation reagents are commercially available; common reagents are shown in Table 1.1.

Table 1.1 Commonly available biotinylation reagents

Reagent	Structure	Specificity	Water Solubility	Cleavability
Sulfo-NHS-Biotin M.W. 443.42 13.5 Å	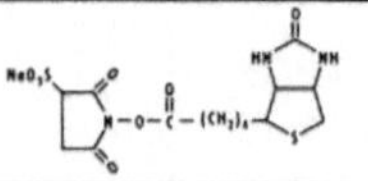	Amines	Soluble	Noncleavable
NHS-Biotin M.W. 341.38 13.5 Å	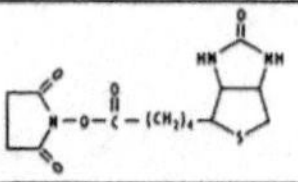	Amines	Insoluble	Noncleavable
NHS-LC-Biotin M.W. 556.58 22.4 Å	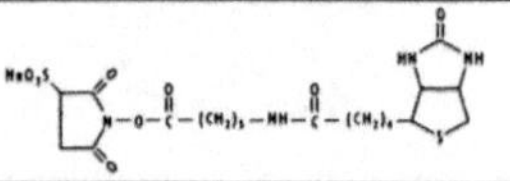	Amines	Soluble	Noncleavable
NHS-LC-Biotin II M.W. 454.54 22.4 Å	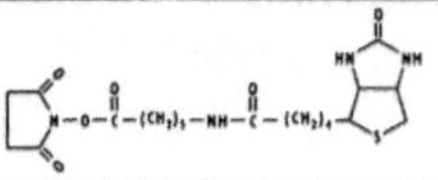	Amines	Insoluble	Noncleavable
NHS-SS-Biotin M.W. 606.70 24.3 Å	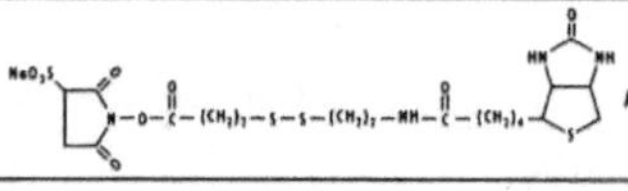	Amines	Soluble	Thiol cleavable
NHS-Iminobiotin M.W. 421.32 13.5 Å	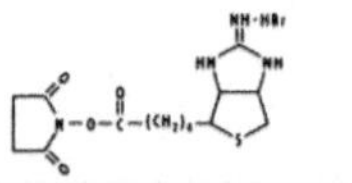	Amines	Insoluble	Noncleavable
Biotin Hydrazide M.W. 258.34 15.7 Å	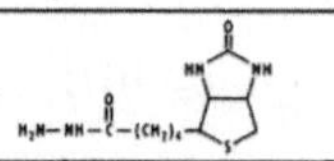	Carbohydrates and Carboxyls	Soluble	Noncleavable
Biotin-LC-Hydrazide M.W. 371.5 24.7 Å	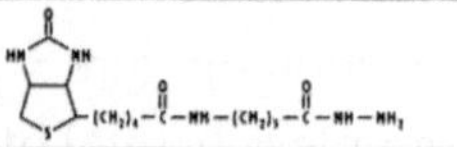	Carbohydrates and Carboxyls	Soluble	Noncleavable
Biocytin Hydrazide M.W. 386.51 19.7 Å	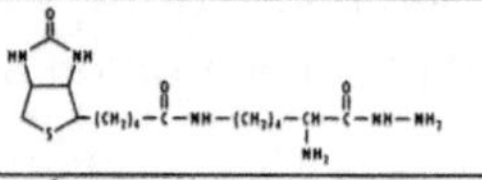	Carbohydrates and Carboxyls	Soluble	Noncleavable
5-(Biotinamido)pentylamine M.W. 328.48 20.0 Å	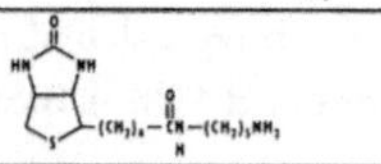	Carboxyls	Soluble	Noncleavable
Biotin-HPDP M.W. 539.77 29.2 Å	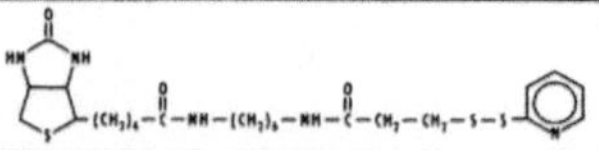	Sulfhydryls	Insoluble	Thiol cleavable
Iodoacetyl-LC-Biotin M.W. 510.42 27.1 Å	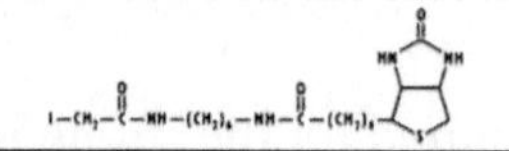	Sulfhydryls	Insoluble	Noncleavable
Biotin-BMCC M.W. 533.69 32.6 Å	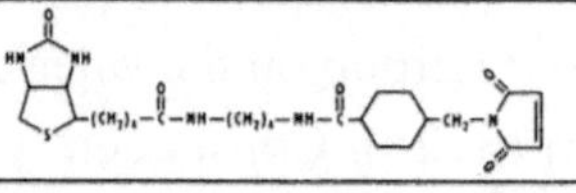	Sulfhydryls	Insoluble	Noncleavable
Photoactivatable Biotin M.W. 533.65 30.0 Å	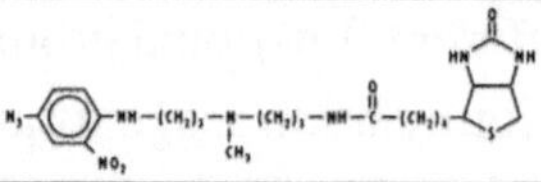	Nonspecific	Insoluble	Noncleavable

The angstrom length indicates the distance imparted by the reagent, from the biotin ring to the reacted group, after reaction with its target group.

General Considerations for Biotinylation

The availability of biotinylation reagents having different functional group reactivities allows the researcher tremendous flexibility. A biotinylation reagent can be chosen to react with a given group of molecules or to exclude a reaction with a given molecule. The ability to choose a biotinylation reagent that reacts specifically to a particular functional group often facilitates preservation of the inherent biological activity of a compound. An informed decision about the choice of biotinylation reagent can often be made by drawing on other available information. For example, if a critical amino acid residue(s) in the binding site of a receptor is known, a biotinylation reagent that does not react with this residue(s) can be chosen. Similarly, if FITC labeling (an amine-reactive chemistry) has destroyed the ability of a protein to bind to its receptor, there is a good probability that an amine-reactive biotinylation reagent would also inhibit the ability of the protein to bind to its receptor.

In the case of proteins, only functional groups on the surface of the protein's tertiary structure will be biotinylated, since the protein is not normally denatured under the conditions of biotinylation. With proteins of any significant size, it would be the rare exception were neither carboxyl groups nor primary amine groups available, but in the case of cysteines a good proportion of proteins will lack cysteine (in the reduced form) on the surface. When inferring information about a protein from an amino acid sequence, it does not necessarily follow that presence of cysteine in the sequence would give rise to an available sulfhydryl group on the surface that can be targeted by a sulfhydryl-reactive biotinylation reagent. The use of Ellman's reagent can be used to confirm the presence of an available sulfhydryl group on a protein (26, 27).

Proteins can often be multiply biotinylated, especially with the use of amine- and carboxyl-reactive biotinylation reagents. Especially with respect to applications involving interacting biological partners (i.e., ligand/receptor or antigen/antibody), the "best" degree of biotinylation often represents an interplay between higher biotin incorporation and loss of biological activity. For a particular protein that has been biotinylated to the extent of 2.5 mol of biotin per mol of protein, for example, one can envision a Gaussian (bell-shaped) distribution of biotin incorporation in the protein pool. While some proteins in the pool may not have incorporated any biotin, most would have incorporated 2–3 mol, and a very small fraction of the protein pool 5 mol. The range of biotin distribution in the pool can be critical for some applications. An overbiotinylation effect may be seen in some instances where a protein is inactive at 5 mol of biotin incorporation, but at an average of 4 mol of biotin incorporation the protein retains 15–20% of its activity. If the range of biotin distribution is critical, particular attention should be given to the details of the reaction, such as the amount of biotinylation reagent, reaction time, and so on. The HABA assay for biotin can be particularly useful in optimizing the reaction to give the "best" biotinylated product. Alternatively, a molecule could be biotinylated under a given set of conditions and then separated by chromatography techniques such as reversed-phase HPLC or ion exchange to arrive at the "best" fractions for the purpose at hand. Immobilized monomeric avidin, which has a weaker affinity for biotin compared with tetrameric avidin, can be used under physiologically compatible conditions to remove nonbiotinylated material from a preparation.

While it is easy to envision random biotinylation of a protein, this is not necessarily the case. In the case of amine-reactive biotinylation reagents, if each epsilon amine on the surface of the protein had an identical pK', it would follow that the distribution of biotin on the surface of the protein should also be Gaussian. The actual distribution can be affected by changes in the pK' of the epsilon amine due to the microenvironment in which they reside. Other microenvironmental features that can affect the distribution and/or incorporation of biotin include the relative hydrophobicity/hydrophilicity of the site and steric hindrance caused by neighboring groups. The choice of water-soluble versus water-insoluble biotinylation reagents, short-chain versus long-chain reagents, and specific reactive groups as in the case of the sulfhydryl-reactive reagents, can occasionally have a profound impact on the utility of the biotinylated compound because of these microenvironmental effects. Generally, however, water-soluble biotinylation reagents and biotinylation reagents with long-chain spacer arms are usually preferred due to their greater ease of use and typically higher detectability.

Amine-Reactive Biotinylation Reagents

NHS (*N*-hydroxysuccinimide) esters of biotin are the most commonly used biotinylation reagents. Several different NHS esters of biotin are available, each with varying spacer arm lengths and other properties such as cleavability or reversibility of binding to avidin or streptavidin. The NHS esters can be broadly divided into 2 separate classes, the water-soluble and water-insoluble. The reactive chemistry of these two classes of esters toward primary amines is essentially identical, and is illustrated in Fig. 1.2. As implied by their class, the water-soluble NHS esters can be directly used in aqueous reaction mixtures. Water-insoluble NHS esters are first dissolved in an organic solvent, most commonly dimethyl sulfoxide (DMSO), and then aliquoted into an aqueous reaction mixture with a solvent carryover of up to 10%. The preparation of stock solutions with the intent of long-term storage is not recommended for either class of compounds, since they hydrolyze easily. Even in the case of water-insoluble NHS esters, the solvents used to dissolve them (DMSO or DMF, dimethylformamide) are hygroscopic and thus tend to absorb water and promote hydrolysis of the NHS ester.

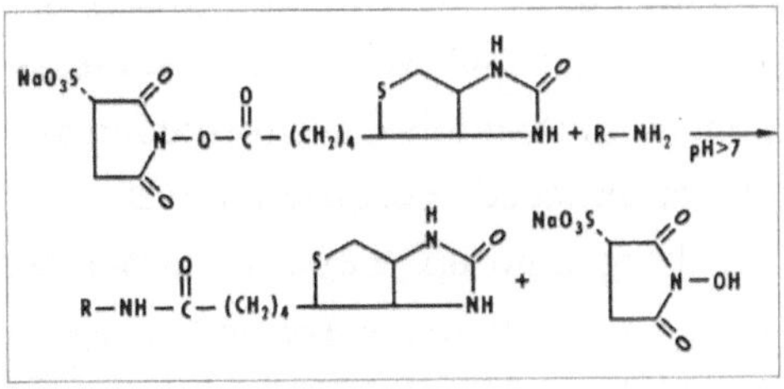

Figure 1.2 Reaction of NHS and sulfo-NHS esters of biotin with a primary amine

Water-Soluble NHS Esters of Biotin

Sulfo-NHS-biotin, sulfo-NHS-LC-biotin, and NHS-SS-biotin are water-soluble NHS esters of biotin. Their water-soluble characteristics are made possible by the presence of the sulfonate ($-SO_3$) group on the *N*-hydroxysuccinimide ring. The water solubility of these compounds is an extreme advantage in terms of convenience (i.e., no need to first dissolve in a solvent) and the solvent incapability of some applications. Additionally, the charged sulfonate group imparts membrane impermeability to these reagents, making them useful for cell surface-labeling protocols. Sulfo-NHS-biotin and NHS-LC-biotin are noncleavable biotinylation reagents, in that the biotin moiety cannot be removed from the biotinylated molecule under normal conditions. NHS-SS-biotin is unique from the standpoint that it contains a disulfide bond in its spacer arm, making the biotin moiety cleavable after biotinylation reactions with reducing agents such as 50 mM DTT, 100 mM β-mercaptoethanol, or 1% sodium borohydride.

These sulfonated NHS esters of biotin are supplied as their sodium salts and are soluble in water to at least a concentration of 10 mM. However, in the presence of high concentrations of cations (i.e., due to high salt buffers), these compounds may begin to precipitate, or "salt out", from solution. Methods to overcome problems with solubility include simple dilution of the reaction system with water and changing the buffer composition.

Water-Insoluble NHS Esters of Biotin

NHS-biotin, NHS-LC-biotin II, and NHS-iminobiotin are water-insoluble NHS esters of biotin. These compounds are capable of biotinylation in aqueous solutions if they are first dissolved in an organic solvent, then aliquoted into the aqueous reaction mixture. The most commonly used solvent is DMSO. When these solvents are used as carriers for the biotinylation reagent, a microemulsion is formed in the aqueous phase, and the biotinylation reaction can proceed. Because the water-insoluble NHS esters of biotin do not possess a charged group, they are membrane permeable to cells. NHS-biotin and NHS-LC-biotin II are noncleavable biotinylation reagents. NHS-iminobiotin is unique from the standpoint that it exhibits pH-dependent binding characteristics to both streptavidin and avidin (28). At pH 9.5 or greater, avidin will bind to the iminobiotin moiety tightly. Complete dissociation of the avidin-iminobiotin complex occurs at pH 4. NHS-iminobiotin can be used for applications that require mild dissociation conditions from avidin or streptavidin, or in cases where the application cannot tolerate the reducing conditions required to break the disulfide bond of reagents such as NHS-SS-biotin.

Biotinylation reactions with the water-insoluble NHS esters are typically carried out with a solvent carryover of up to 10% final volume in the aqueous reaction mixture. These reagents can begin to fall out of the aqueous solution at high concentrations, as noted by the appearance of a milky turbid solution. While the biotinylation reaction will still take place under such conditions, a protocol can be modified to ensure complete dissolution of the NHS ester. For example, the aqueous phase can be supplemented with additional organic solvent. DMSO gives better solubility to these reagents than does DMF.

General Considerations for the Use of NHS-Ester Biotinylation Reagents

Two considerations are crucial to the use of these reagents: the specificity and definition of the amine reactivity, and the effect of pH on the reaction and reagent. Aliphatic primary amines are the principal targets for NHS esters. Aromatic amines present on nucleic acid bases are not reactive. Alpha amine groups, present on the N-terminus of peptides and proteins, can be biotinylated if they can interact with the NHS ester. While 5 amino acids have nitrogen in their side chains, only the epsilon amine of the lysine side chain can be effectively biotinylated with NHS esters (29). The amides present on asparagine and glutamine are not reactive, nor is the guanidino nitrogen of arginine or the cyclic secondary amine of tryptophan. The imidazole group of histidine effectively increases the hydrolysis rate of NHS esters, since it forms an unstable intermediate. Extremely limited reactivity occurs toward hydroxyls on serine and tyrosine due to transesterification. The transesterification products are labile to hydroxylamine treatment or exposure to alkaline pH. Cross-reactivity can occur toward cysteine due to the nucleophilic character of the side chain sulfhydryl.

Primary amines also are found in the structure of Tris, which makes Tris an unacceptable buffer for NHS-ester reactions. However, Tris can be used to advantage in a biotinylation protocol. A large excess of Tris at a neutral to basic pH can be added at the end of an NHS-ester reaction to quickly quench and terminate the reaction. Glycine and ethanolamine can be used in a similar fashion.

Two factors are related to the pH dependence of the NHS-ester reaction. One factor is that because this is a nucleophilic reaction, the rate of reaction with primary amines increases with increasing pH. The other factor is that the rate of hydrolysis of NHS esters also increases with increasing pH. Consequently, an optimum reaction pH for NHS-ester reagents cannot be stated practically since it depends on several other factors, such as amine density on the compound, and actual pK' of the amine(s), concentration, and so on. Typical reaction pH values, however, are 7–9. Studies performed on NHS-ester-containing compounds indicate that the half-life of hydrolysis for NHS esters is 4–5 hours at pH 7.0 in an aqueous environment free of primary amines (29). This half-life decreases to the order of 1 hour at pH 8.0 and 25 °C (30). It further decreases to the order of 10 min at pH 8.6 and 4 °C (31).

Biotinylating IgG with NHS-Ester Biotinylation Reagents

A biotinylated IgG, with a degree of insertion at approximately 2 mol of biotin per mol of IgG, can be prepared from either the water-soluble or water-insoluble NHS esters of biotin using the protocol below. The protocol uses an approximate 10-fold molar excess of biotinylation reagent over the amount of IgG. Other buffers such as PBS, HEPES, and borate can also be used in the protocol. The reaction time should be extended to 2 hours when using physiological pH buffers. While this protocol can also be used for NHS-iminobiotin, the use of IgG limits potential applications of the biotinylated product. The iminobiotin moiety requires high pH for effective binding to avidin or streptavidin, conditions which would abolish most antigen/antibody binding.

Materials

Coupling buffer. Composition is 50 mM bicarbonate buffer, pH 8.5.

IgG solution. Prepare at 2 mg/ml in coupling buffer.

Biotinylation reagent stock solution. Prepare immediately before use. For the water-soluble NHS-ester reagents, prepare by adding 1 mg of the reagent to 1 ml of distilled water. For the water-insoluble NHS ester reagents, prepare by adding the reagent to a 13.3 mM final concentration in DMSO (4.5 mg of NHS-biotin/ml of DMSO or 6 mg of NHS-LC-biotin II/ml of DMSO).

Quench buffer. 1 M Tris, pH 8.5. The use of a quench buffer is optional in this protocol, but it can be particularly useful when the intent is to vary the reaction conditions (i.e., molar excess of reagent) to determine the best biotinylated product for the application at hand.

Protocol 1.2

NHS-Ester-Biotinylation Reaction

1. Add the biotinylation reagent stock solution to the IgG solution. For the water-soluble reagents, add 74 µl of an NHS-LC-biotin stock per ml of IgG solution; add 60 µl of an NHS-biotin stock per ml of IgG solution; add 80 µl of an NHS-SS-biotin stock per ml of IgG solution. For the water-insoluble reagents, add 10 µl of the DMSO stock solution per ml of IgG solution.
2. Mix briefly and react for 30 min at room temperature.
3. Optionally, the reaction can be quenched by the addition of Tris, pH 8.5, to a final concentration of 50 mM.
4. Excess and hydrolyzed biotinylation reagent can be removed by gel filtration, dialysis, or with microconcentrators.

Carbohydrate-Directed Biotinylation Reagents

Biotin hydrazide, biotin-LC-hydrazide, and biocytin hydrazide are biotinylation reagents that can be used to target carbohydrate groups on macromolecules. Oxidative pretreatment of glycoproteins is used to generate reactive aldehyde groups that couple the hydrazide ($R\text{-}NHNH_2$) moiety to form a hydrazone linkage. This reaction is shown in Fig. 1.3. The hydrazone bond formed in this reaction is stable between pH 2 and 10 (35). This approach can be used with any glycoprotein that contains a *cis*-diol in its carbohydrate structure. Hydrazide derivatives of biotin are particularly useful for biotinylating antibodies, since the biotinylation reaction is predominately localized to the Fc region of the immunoglobulin, away from its antigen-binding site (36).

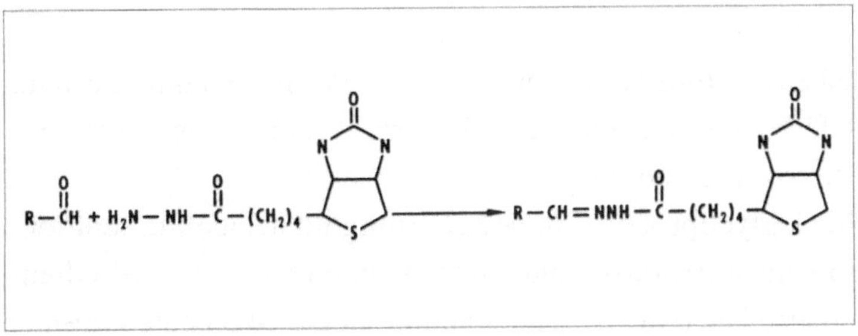

$$R-\overset{O}{\overset{\|}{C}}H + H_2N-NH-\overset{O}{\overset{\|}{C}}-(CH_2)_4-\text{[biotin ring]} \longrightarrow R-CH=NNH-\overset{O}{\overset{\|}{C}}-(CH_2)_4-\text{[biotin ring]}$$

Figure 1.3 Reaction of biotin hydrazide with an aldehyde *Cis-Diols of the carbohydrate are previously oxidized to an aldehyde, which reacts with the hydrazide moiety.*

Considerations for the Use of Hydrazide Derivatives of Biotin

Sodium *m*-periodate ($NaIO_4$) at 10 mM concentration can be used to label carbohydrates in general. Sialic acid residues on glycoproteins can be specifically oxidized with periodate under controlled conditions (37). At 1 mM $NaIO_4$, and a temperature of 0 °C, the reaction is restricted primarily to sialic acid residues. Sialic acid residues can also be biotinylated with hydrazide derivatives by treatment with neuraminadase to generate galactose groups. The galactose and *N*-acetylgalactosamine residues on whole cells can be selectively biotinylated with biotin hydrazides by pretreating the cells with galactose oxidase (38–41). This enzyme will convert the primary hydroxyl groups on these sugars to their corresponding aldehydes.

Biotin hydrazide and biotin-LC-hydrazide are soluble in aqueous buffers up to a concentration of approximately 5 mM. They can also be first dissolved into DMSO (at concentrations of up to 50 mM), then aliquoted into aqueous reaction mixtures. This would be analogous to the techniques used for the water-insoluble NHS esters of biotin. However, biotin hydrazide and biotin-LC-hydrazide are essentially insoluble in DMF (less than 5 mM). Biocytin hydrazide was introduced due to its enhanced water-solubile characteristics.

Since the temperature and pH of oxidation, along with the concentration of periodate, affect the specifics of the reaction, and glycosylation varies with each protein, optimum conditions for each glycoprotein will need to be determined. Each particular protein preparation will have an optimum pH for oxidation and an optimum pH for the hydrazide biotinylation. The hydrazide reaction has been discussed with regard to the differential effects of oxidation and reaction pH conditions with goat and rabbit antibodies (35, 37). Tris, or other primary amine-containing buffers, are not recommended for use in either the oxidation or biotinylation steps; the bond formed between the primary amine and the aldehyde (a Schiff base) would quench the reaction in the presence of the biotin hydrazide.

Materials

Labeling solution. Composition is 100 mM sodium acetate, 0.02% sodium azide, pH 5.5. Store at room temperature and use within 2 months.

Stock glycoprotein. Prepare at 0.1 mg/ml in labeling solution.

Sodium *m*-periodate solution for glycoproteins: 30 mM sodium *m*-periodate ($NaIO_4$). This solution should be prepared fresh.

Sodium *m*-periodate solution for preferential biotinylation of sialic acid residues: 3 mM $NaIO_4$. This solution should be prepared fresh.

Biotin-hydrazide solution: 5 mM biotin hydrazide or biotin-LC-hydrazide. Prepare by dissolving the required amount in labeling solution. Store at 4 °C and use within 2 months.

Stop solution: 0.1 M Tris, pH 7.5.

Protocol 1.3

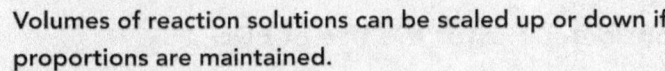

Biotinylation with Hydrazide Derivatives

1. Dissolve glycoprotein (0.1 mg/ml) in labeling solution.
2. Pipette 20 µl of the 0.1 mg/ml solution from step 1 into a 1.5-ml microfuge tube.

Volumes of reaction solutions can be scaled up or down if proportions are maintained.

3. Add 10 µl of 30 mM sodium m-periodate solution to the tube in step 2.
4. Incubate in the dark for 30 min at room temperature to produce aldehyde groups on the carbohydrate portions of the glycoprotein. Note: Sialic acid residues can be preferentially biotinylated using 3 mM $NaIO_4$ at 0 °C for 30 min in the dark.
5. Run the reaction mixture through a gel filtration column to remove $NaIO_4$.

Do not use a carbohydrate-based media such as agarose or Sephadex, because the periodate can oxidize the support and cause retention of oxidized protein; a polyacrylamide-based medium is recommended.

6. Add 10 µl of 5 mM biotin hydrazide.
7. Incubate for 1 hour at room temperature to form biotinylated glycoprotein.
8. Terminate the reaction by adding 50 µl of stop solution. This step quenches the amine reactivity of any remaining aldehydes on the oxidized glycoprotein.
9. Unreacted biotin hydrazide can be removed by gel filtration, dialysis, or with microconcentrators.

Carboxyl-Reactive Biotinylation Reagents

EDC, or 1-ethyl-3-(3-dimethylaminopropyl)-carbodiimide hydrochloride, a water-soluble carbodiimide, can be used to couple 5-(biotinamido)pentylamine to a carboxyl group of a macromolecule, forming an amide linkage that is quite stable (Fig. 1.4) (42). EDC can also be used in a similar fashion with biotin-hydrazide compounds, but the resultant bond is substantially less stable (43).

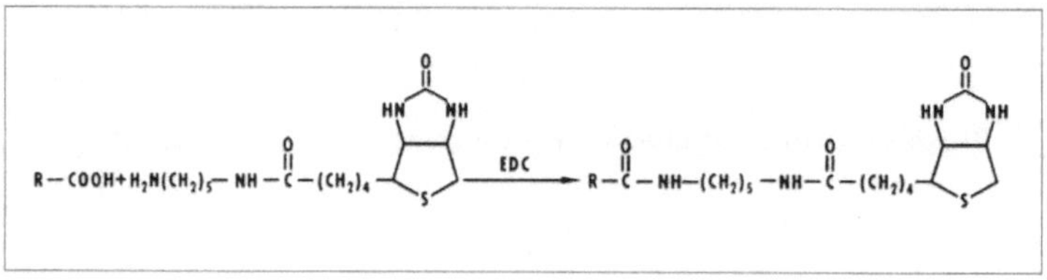

$$R{-}COOH + H_2N(CH_2)_5{-}NH{-}\overset{O}{\overset{\|}{C}}{-}(CH_2)_4{-}\text{[biotin: HN, NH, O, S]} \xrightarrow{EDC} R{-}\overset{O}{\overset{\|}{C}}{-}NH{-}(CH_2)_5{-}NH{-}\overset{O}{\overset{\|}{C}}{-}(CH_2)_4{-}\text{[biotin: HN, NH, O, S]}$$

Figure 1.4 Reaction of 5 (biotinamido)-pentylamine with a carboxyl group using EDC

Considerations for the Use of EDC

The use of EDC may result in some polymerization of the macromolecule if it has both carboxyl groups and primary amines on its surface. If such is the case, the extent of polymerization can be minimized by decreasing the amount of EDC in the reaction and/or increasing the amount of the 5-(biotinamido)pentylamine used in the reaction.

The protocol below relies on the use of a large excess of 5-(biotinamido)pentylamine to avoid excessive polymerization of the protein. Polymerization will not occur if the macromolecule to be reacted lacks amine groups or has had the amine groups previously blocked.

Materials

MES buffer. Composition is 0.1 M MES (*N*-morpholinoethane sulfonic acid), pH 5.5.

5-(Biotinamido)pentylamine. May be used as a dry powder or prepared as a stock solution in MES buffer, which can be frozen long-term.

EDC solution. Prepare immediately before use by dissolving 100 mg of EDC per ml of MES buffer.

Protocol 1.4

Coupling via EDC

1. Dissolve the protein in MES buffer to a concentration of 5–10 mg/ml.
2. Add 5-(biotinamido)pentylamine to a final concentration of 5 mM.
3. Add 12.5 µl of the EDC solution to 1 ml of the protein solution and mix.
4. Incubate overnight at room temperature with stirring.
5. Remove by centrifugation any precipitate that formed during the reaction.
6. Excess reagent and hydrolyzed EDC can be removed by gel filtration, dialysis, or with microconcentrators.

Sulfhydryl-Reactive Biotinylation Reagents

Iodoacetyl-LC-biotin, biotin-HPDP, and biotin-BMCC are biotin derivatives that are reactive toward sulfhydryl groups. The biotin residue imparted by biotin-HPDP can be cleaved away from the reacted sulfhydryl to regenerate the protein (or peptide) into its original, unmodified form. Additionally, the reaction of biotin-HPDP with thiols (Fig. 1.5) can be easily followed with spectrophotometry, because the reaction's leaving group, pyridine-2-thione, has a characteristic absorption maximum at 343 nm with an extinction coefficient of 8.08 +/– 0.3 × 10^3 M^{-1} cm^{-1} (45). Iodoacetyl-LC-biotin is a noncleavable reagent that reacts mainly with thiol groups at pH 7.5–8.5. The reaction occurs by nucleophilic substitution of iodine with a thiol group, resulting in a stable thioether bond (Fig. 1.6). Biotin-BMCC is a noncleavable reagent that reacts with sulfhydryls between pH 6.5 and 7.5 to form a thioether linkage (Fig. 1.7).

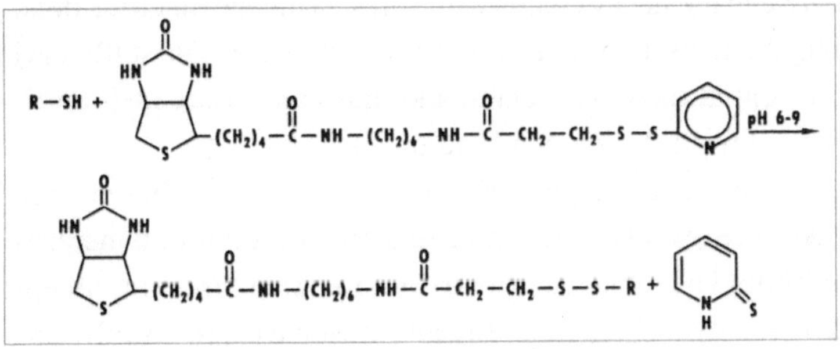

Figure 1.5 Reaction of biotin-HPDP with a sulfhydryl

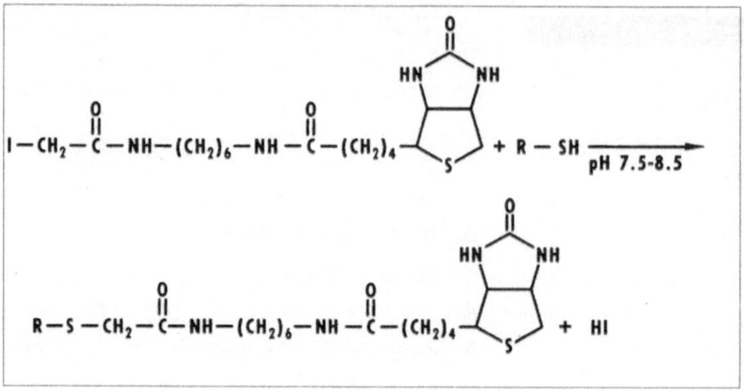

Figure 1.6 Reaction of iodoacetyl-LC-biotin with a sulfhydryl

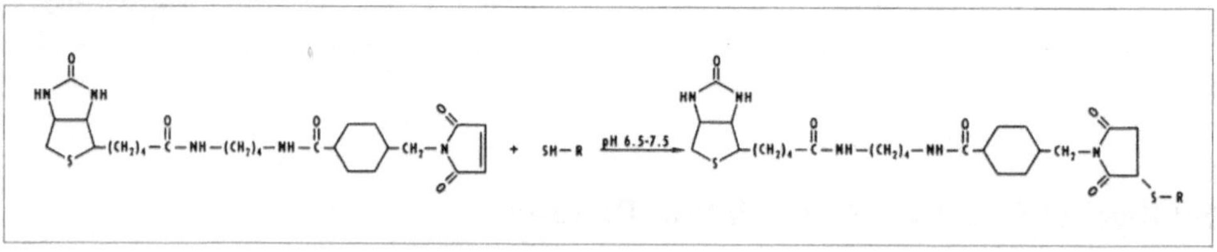

Figure 1.7 Reaction of biotin-BMCC with a sulfhydryl

Considerations for the Use of Sulfhydryl-Reactive Biotinylation Reagents

Most commonly, the source of the sulfhydryl group for biotinylation is the cysteine on the surface of a protein or within a peptide. If a sulfhydryl group is not available, it can be introduced through prior reaction of the macromolecule with the use of SATA (46), Traut's reagent (47), or SPDP (32). Biotinylation of sulfhydryl groups often provides an advantage in an application. For example, a protein in a complex mixture can be targeted for biotinylation if it is the only one with a free sulfhydryl group on its surface. IgG can be reduced under mild conditions, breaking the disulfide bond between the heavy chains while preserving the disulfide bond between the heavy and light chains, to form a monovalent fragment. The sulfhydryl of the reduced IgG is present at the hinge region and thus can be targeted for biotinylation at a site removed from the antigen-binding site.

Reactions with these biotinylation reagents should be carried out in buffers free of extraneous thiols. Therefore, excess β-mercaptoethanol, dithiothreitol, mercaptoethylamine, and so on should be removed prior to biotinylation. Proteins or peptides to be biotinylated by sulfhydryl-reactive reagents must have a free sulfhydryl group that can be confirmed by the use of Ellman's reagent. Disulfides will not be reactive toward sulfhydryl-specific biotinylation reagents, so these must first be reduced and then desalted to remove excess reducing agent. EDTA is particularly useful for inclusion in these protocols due to its antioxidative effect. In a study with reduced IgG fragments, thiol groups were substantially stabilized by the presence of EDTA (48). In 0.1 M sodium phosphate buffer at pH 6–7, 4 °C, the number of free thiols decreased 63–90% and 15–25% in the absence of EDTA with reduced IgG and reduced $F(ab')_2$, respectively. The number of free thiols decreased only 7% and 9% in the presence of EDTA over a 40-hour incubation. The use of nitrogen-sparged buffers is an additional precaution that can be used to prevent reoxidation of free sulfhydryls to disulfides.

Biotinylation of Surface Thiols with Biotin-HPDP

Protocol 1.5 illustrates the use of Biotin-HPDP for targeting surface sulfhydryl groups on proteins. The method is illustrated with *β*-D-galactosidase, a protein that contains surface sulfhydryl groups. Consequently, the method does not require pretreatment with a reducing agent. This protocol can be adapted as necessary for specific applications. Biotin-HPDP can be dissolved in a variety of dipolar aprotic solvents, with DMSO a typical choice. The biotinylation reaction can be carried out in a variety of buffers and at a pH range of 6–9. In addition to using dithiothreitol, biotin can also be cleaved away from the biotinylated compound by using 100 mM *β*-mercaptoethanol or 1% sodium borohydride. Because sodium borohydride is a very strong reducing agent, it may destroy functional groups on proteins or peptides.

Materials

4 mM Biotin-HPDP stock solution: Prepare by adding 2.2 mg of biotin-HPDP to 1.0 ml of DMF. This stock can be aliquoted and stored frozen.

Phosphate-buffered saline/EDTA buffer (PBS-EDTA): The composition of this buffer is 20 mM sodium phosphate, 150 mM NaCl, 1 mM EDTA, pH 7.4.

β-D-galactosidase

Protocol 1.5

Biotinylation of Surface Thiols with Biotin-HPDP

1. Dissolve 4 mg of β-D-galactosidase in 2.0 ml of PBS-EDTA.
2. Add 0.1 ml of the biotin-HPDP stock solution to 2.0 ml of the β-D-galactosidase solution.
3. Mix and incubate for 90 min at room temperature.
   The progress of the reaction can be monitored by following the change in absorbance at 343 nm due to the release of pyr dine-2-thione.
4. Excess reagent can be removed by gel filtration, dialysis, or with microconcentrators.

Biotinylating Reduced IgG with Iodoacetyl-LC-Biotin

The protocol below uses 2-mercaptoethylamine·HCl, a weak reductant, to selectively reduce the disulfides between the heavy chains of an IgG while preserving the disulfide bonds between the heavy and light chains. Generation of a thiol in the hinge region of the monovalent antibody directs the biotinylation away from the antigen-binding site. The reduction with MEA can be carried out between pH 6.0 and 9.0. IgG concentrations between 1 and 20 mg/ml can be effectively reduced with a mercaptoethylamine HCl concentration of approximately 50 mM.

Materials

Iodoacetyl-LC-biotin stock solution: Prepare at 4 mM in DMF.
Phosphate/EDTA buffer: Composition is 0.1 M sodium phosphate, 5 mM EDTA, pH 6.0.
Tris/EDTA buffer: Composition is 50 mM Tris, 5 mM EDTA, pH 8.3.
Mercaptoethylamine-HCl: Used to reduce IgG.

Protocol 1.6

Biotinylating Reduced IgG with Iodoacetyl-LC-Biotin

1. Dissolve 4 mg of IgG in 200 µl of phosphate/EDTA buffer.
2. Add 1.4 mg of mercaptoethylamine-HCl, mix and incubate at 37 °C for 90 min.
3. Cool the solution to room temperature and perform a buffer exchange to remove excess mercaptoethylamine. Apply the mixture to a desalting column that has been preequilibrated with Tris/EDTA buffer.
4. Collect 0.5-ml fractions. Monitor absorbance at 280 nm and pool fractions containing peak absorbances.
5. Add 15 µl of iodoacetyl-LC-biotin stock solution to 1 ml of the reduced IgG from step 4 above.
6. Mix well and incubate in the dark for 90 min at room temperature. [This limits conversion of liberated iodide ion to molecular iodine, which could react with tyrosine residues (49).]
7. Excess reagent can be removed by gel filtration, dialysis, or with microconcentrators.

Biotinylating Reduced IgG with Biotin-BMCC

Protocol 1.7 illustrates the use of biotin-BMCC for biotinylating reduced IgG. The reaction can be modified to include other buffers between pH 6.5 and 7.5. DMF can be substituted for DMSO for the preparation of the stock solution, but the solubility of the compound is greatly reduced in DMF (approximately 33 mM in DMSO versus approximately 7 mM in DMF).

Materials

Biotin-BMCC stock solution. Prepare at 8.5 mM in DMSO.
Phosphate/EDTA buffer. Composition is 0.1 M sodium phosphate, 5 mM EDTA, pH 6.0.
PBS/EDTA buffer. Composition is 0.1 M sodium phosphate, 0.15 M NaCl, 5 mM EDTA, pH 7.2.
2-Mercaptoethylamine·HCl

Protocol 1.7

Biotinylating Reduced IgG with Biotin-BMCC

1. Dissolve 5.0 mg of IgG in 2 ml of phosphate/EDTA buffer.
2. Add 6 mg of 2-mercaptoethylamine·HCl, mix, and incubate at 37 °C for 90 min.
3. Cool the solution to room temperature and perform a buffer exchange to remove excess mercaptoethylamine. Apply the mixture to a desalting column that has been preequilibrated with PBS/EDTA buffer.
4. Collect 1-ml fractions. Monitor absorbance at 280 nm, and pool fractions containing peak absorbances. Adjust protein concentration to 1 mg/ml.
5. Add 100 µl of biotin-BMCC stock solution to 2.5 ml of the reduced IgG from step 4 above.
6. React for 2 hours at room temperature.
7. Excess reagent can be removed by gel filtration, dialysis, or with microconcentrators.

Phenylazide Derivatives of Biotin

Phenylazide derivatives of biotin can be used to label proteins, nucleic acids, and other organic molecules in which primary amines, carboxyls, sulfhydryls, or carbohydrate groups are not readily available or may be necessary for functional reasons. Photoactivatable biotin is an example of this class of reagents. It reacts nonspecifically on photoactivation via a reactive nitrene intermediate. The nitrene that results from photoactivatable biotin on photolysis also can react with C–H bonds (52), although it is preferential for nucleophiles in heterologous systems (53). The nitro ($-NO_2$) substitution on the phenylazide group enables the compound to be activated at 350 nm, eliminating the need for light at lower wavelengths (i.e., 254 nm), which may promote thymine dimerization with oligonucleotides. Photoactivatable biotin can be used for labeling double-stranded DNA and single-stranded DNA or RNA. Single-stranded DNA is not degraded or cross-linked by the labeling protocol (54). Labeling can be monitored visually, because the biotinylated nucleic acid is red (54).

Considerations for the Use of Photoactivatable Biotin

The azide group on the reactive phenyl ring of the compound is a pseudohalogen and can be hydrolyzed, under acidic conditions, to hydrazoic acid. Exposure of photoactivatable biotin to acid pH before photoactivation is not recommended, therefore, since it would lead to loss of reactivity of the compound. Phenylazide reagents also have limited stability to reducing agents prior to photoactivation.

Biotinylation with Photoactivatable Biotin

Protocol 1.8 illustrates the use of photoactivatable biotin for biotinylating a nucleic acid probe. When the nucleic acid is biotinylated as recommended, a ratio of 1 biotin to 100–400 residues results (54, 55). Modification at this level decreases the probability of interference with the recognition of complementary sequences by the probe.

Materials

Photoactivatable biotin stock solution: 1 mg/ml in distilled water. This stock solution can be stored, protected from light, at –20 °C for up to 1 year.

Nucleic acid probe solution. Prepare at a concentration of approximately 1 µg/ml in water or 0.1 mM EDTA, pH 8.0.

Precipitation reagents. 2-Butanol, 4.0 M NaCl, ethanol, and dry ice.

Light source with appropriate emission wavelength energy, such as high-intensity mercury vapor lamp, 250 W. Other light sources can also be used, such as handheld long-wave UV lamps, or camera flash units that mimic a daylight spectrum. Irradiation times (or number of flashes) will need to be adjusted to ensure an adequate reaction. Care should be taken that the sample is not overheated by the light source.

Protocol 1.8

Biotinylation of Nucleic Acids

1. In a microfuge tube, mix equal volumes of nucleic acid probe and photoactivatable biotin stock solution.
2. Incubate in an ice bath 10 cm below the lamp for 15 min. Shine the light down through the open microfuge tube.
3. Increase the sample volume to 100 µl by the addition of 0.1 M Tris-HCl, pH 9.0.
4. Add 100 µl of 2-butanol. Mix sample and centrifuge.
5. Discard the upper phase.
6. Add another 100 µl of 2-butanol. Mix.
7. Discard upper phase.
8. Add 10–50 µg of unrelated carrier DNA or RNA.
9. Add 0.75 µl of 4.0 M NaCl. Mix.
10. Add 100 µl of ethanol for DNA or 125 µl of ethanol for RNA.
11. Cool sample on dry ice for 15 minutes or at –20 °C overnight.
12. Collect precipitate by centrifugation.
13. Dry sample and dissolve in 0.1 mM EDTA, pH 8.0. Biotinylated nucleic acid probes prepared in this manner can be stored at –20 °C for at least a year. These biotinylated probes can be used in hybridization techniques in the same manner as biotinylated probes prepared by other methods.

1.3 Troubleshooting

Avidin-HABA Biotin Assay

The avidin-HABA reagent will have an initial A_{500} higher than 0.9 if higher concentrations of avidin are used in the reagent's preparation. Dilutions must be accounted for during calculations. Substantial decreases in the final pH of the solution caused by the use of highly buffered acidic solutions will cause a decrease in the extinction of the avidin-HABA reagent and result in underestimation of the biotin concentration of the sample. Always run a control consisting of the nonbiotinylated compound; a few proteins will bind HABA dye (causing an increase in the A_{500}) and result in underestimation of biotin incorporation. Pronase digestion can often be used to prevent HABA from binding to these proteins.

NHS-Ester Biotinylation Reactions

Protocol 1.2 can be used with very little difficulty with polyclonal antibodies. Several exceptions are seen with monoclonal antibodies where loss of activity occurs due to modification of the antigen-binding sites; a protocol involving a biotin hydrazide compound is recommended in these instances. The leaving group in the NHS-ester reaction (NHS or sulfo-NHS) can cause problems with protein determination. NHS and sulfo-NHS can contribute strong A_{280} absorbance to the protein pool (32–34), especially at alkaline pH, and will also contribute to color formation in the BCA protein assay. Coomassie-based protein assays are compatible with NHS and sulfo-NHS. Biotinylations resulting in little or no biotin incorporation can often be traced to the presence of interfering amines or reducing agents in the reaction buffer, to improper storage of the biotinylation reagent, or to the use of buffers with too little buffering capacity. An often overlooked source of amine contamination comes from the use of HCl in the pH adjustment of the reaction buffer. Ammonia (in the form of ammonium ion) is a common contaminant of HCl and can impact negatively, especially with scaled-down reactions on dilute samples. These reagents are best stored at –20 °C over a desiccant; the bottle should be allowed to warm to room temperature for at least 30 min to avoid moisture condensation on the cold powder. Due to the release of protons, poorly buffered reactions will exhibit a decrease in pH as the reaction progresses and thereby slow down the reaction rate. Differential results on sequential preparations can often be traced to varying sample concentration, even though the molar excess of biotinylation reagent is maintained; more dilute samples will require higher molar excesses of biotinylation reagent to compensate for greater relative hydrolysis of the reagent. The water-insoluble NHS esters of biotin will precipitate from solution if the carryover concentration of DMSO is too low.

Biotin-Hydrazide Biotinylation Reactions

Hydrazide approaches typically result in lower levels of biotin incorporation than can be obtained with other chemistries, especially NHS-ester chemistries. The most common problem associated with this protocol comes from the lack of carbohydrate on the protein. This is especially true with monoclonal antibodies, which in the author's experience can result in an inadequate product up to 40% of the time.

The second problem arises from the polymerization of the oxidized protein, which can occur in the protocol if it is stored prior to addition of the biotin-hydrazide compound or if it is exposed to alkaline pH in the gel-filtration step to remove the excess periodate. Polymerization occurs through Schiff-base formation with primary amines on the protein; this process is promoted by alkaline pH.

EDC/5-(Biotinamido)pentylamine Reactions

If a nonpolymerized product is essential, separation of the product by molecular weight-sizing methods is suggested. Thiol-reducing agents such as DTT and β-mercaptoethanol will quickly quench the EDC (42). Phosphate buffers will decrease the efficiency of the EDC reaction (44).

Iodoacetyl-LC-Biotin Reactions

Potential cross-reactivity toward functional groups other than thiols can be a major problem in this reaction. Iodoacetyl-LC-biotin reacts with sulfhydryl groups most rapidly between pH 7.5 and 8.5. The reaction can be directed toward sulfhydryl groups by limiting the molar proportion of iodoacetyl-LC-biotin to protein, such that the concentration of iodoacetyl-LC-biotin is present at a small excess over the sulfhydryl content. Ellman's reagent can be used to determine the number of –SH groups available on the protein (26, 27). Another way to ensure reactivity with sulfhydryls is by choosing the appropriate pH for the reaction. Below pH 9, the reaction with amines, thioether, and imidazole groups is avoided by restricting the concentration of iodoacetyl-LC-biotin. Thus, to assure that the reaction takes place specifically with sulfhydryls, only a slight stoichiometric excess of iodoacetyl-LC-biotin should be used. Also, maintaining the pH at or near 8.3 ensures the modification of sulfhydryl groups and not amino groups. Histidyl side chains and amino groups react in the unprotonated form and may be observed to take part in reactions above pH 5 and pH 7, respectively, although this reaction is much slower than that for sulfhydryls (50).

The mercaptoethylamine reduction protocol works well with a variety of polyclonal antibodies. It can easily be optimized by varying the time of reduction. Aliquots can be pulled and subjected to nonreducing electrophoresis and the fragments analyzed. Do not boil the sample in the presence of SDS. The user is strongly advised to optimize the reduction when using monoclonal antibodies, since there is extreme variability in the ease (or difficulty) of reduction with monoclonals.

Biotin-BMCC Reactions

As long as the reaction pH is between pH 6.5 and 7.5, biotin-BMCC reactions are not prone to the cross-reactivity problems associated with iodoacetyl-LC-biotin. At pH 7, the maleimide group is 1000-fold more reactive toward a free sulfhydryl than toward an amine (51). At pH values greater than 7.5, reactions with amines become more significant. The separation of mercaptoethylamine from the reduced IgG should be confirmed with the use of Ellman's reagent to avoid quenching the maleimide reaction. As stated previously, the mercaptoethylamine reduction should be optimized when using monoclonal antibodies.

Phenylazide Biotinylation Reactions

Reactions resulting in little or no biotin incorporation can often be traced to insufficient light energy input. Light energy decreases with the inverse square of the distance from the object. An object will be illuminated with one-fourth the energy when the light source is 10 cm away as compared with a distance of 5 cm. Turbid solutions of sample can also cause quenching of light energy.

Acknowledgments

The author expresses his thanks to the Pierce Creative Group for assistance in preparing the graphics. The author, Dean Savage, is a technical assistance representative for Pierce Chemical Company, which has supported Mr Savage in his authorship of a chapter in this manual. However, the information provided in this chapter has not been reviewed for accuracy by Pierce Chemical Company, which makes no representation regarding and assumes no responsibility for, the accuracy or completeness of such information.

References

1 Savage, M.D., Mattson, G., Desai, S., Nielander, G. W., Morgensen, S. and Conklin, E.J. (1992) In: Avidin-Biotin Chemistry: A Handbook, Pierce Chemical Company, Rockford, IL.

2 Wilchek, M. and Bayer, E.A. (1988) *Anal. Biochem.* **171**, 1–32.

3 Green, N.M. (1975) In: Advances in Protein Chemistry, Academic Press, New York.

4 Donovan, J.W. and Ross, K.D. (1973) *Biochemistry* **12**, 512–517.

5 Pai, C.H. and Lichstein, H.C. (1964) *Proc. Soc. Exp. Biol. Med.* **116**, 197–200.

6 Wei, R.-D. and Wright, L.D. (1964) *Proc. Soc. Exp. Biol. Med.* **117**, 341–344.

7 Ross, S.E., Carson, S.D. and Fink, L.M. (1986) *BioTechniques* **4**, 350–354.

8 Green, N.M. (1963) *Biochem.* J. **89**, 609–620.

9 Cuatrecasas, P. and Wilchek, M. (1968) *Biochem. Biophys. Res. Commun.* **33**, 235–246.

10 Bodansky, A. and Bodansky, M. (1970) *Experientia* **26**, 327.

11 Woolley, D.W. and Longsworth, L.G. (1942) *J. Biol. Chem.* **142**, 285–290.

12 Dayhoff, M. O. (1972) In: Atlas of Protein Sequence and Structure, Vol. 5, National Biomedical Research Foundation, Washington, D.C.

13 Green, N.M. (1966) *Biochem. J.* **101**, 774–780.

14 Chaiet, L. and Wolf, F.J. (1964) *Arch. Biochem. Biophys.* **106**, 1–5.

15 Chaiet, L., Miller, T.W., Tausig, F. and Wolf, F. J. (1963) *Antimicrob. Ag. Chemother.* **3**, 28–32.

16 Sano, T. and Cantor, C.R. (1990) *J. Biol. Chem.* **265**, 3369–3373.

17 Gitlin, G., Bayer, E.A. and Wilchek, M. (1988) *Biochem. J.* **250**, 291–294.

18 Merck Index, (1989) 11th Edition, Merck & Co., Rahway, NJ, p. 192.

19 Al-Hakim, A.H. and Hull, R. (1986) *Nucl. Acid Res.* **14**, 9965–9976.

20 Chastain, J.L., Bowers-Komro, D.M. and McCormick, D.B. (1985) *J. Chrom.* **330**, 153–158.

21 Lin, H.J. and Kirsch, J.F. (1977) *Anal. Biochem.* **81**, 442–446.

22 McCormick, D.B. and Roth, J.A. (1970) *Meth. Enzymol.* **18A**, 418–424.

23 Mock, D.M., Langford, G., DuBois, D., Criscimagna, N. and Horowitz, P. (1985) *Anal. Biochem.* **151**, 178–181.

24 Green, N.M. (1965) *Biochem. J.* **94**, 23c–24c.

25 Wilchek, M. and Bayer, E.A. (1988) *Anal. Biochem.* **171**, 1–32.

26 Ellman, G.L. (1959) *Arch. Biochem. Biophys.* **82**, 70–77.

27 Riddles, P.W., Blakeley, R.L. and Zerner, B. (1979) *Anal. Biochem.* **94**, 75–81.

28 Orr, G.A. (1981) *J. Biol. Chem.* **256**, 761–766.

29 Lomant, A.J. and Fairbanks, G. (1976) *J. Mol. Biol.* **104**, 243–261.

30 Staros, J.V. (1988) *Account Chem. Res.* **21**, 435–441.

31 Cuatrecasas, P. and Parikh, I. (1972) *Biochemistry* **11**, 2291–2299.

32 Carlsson, J., Drevin, H. and Axen, R. (1978) *Biochem. J.* **173**, 723–737.

33 Partis, M.D., Griffiths, D.G., Roberts, G.C. and Beechey, R.B. (1983) *J. Prot. Chem.* **2**, 263–277.

34 Abdella, P.M., Smith, P.K. and Royer, G.P. (1979) *Biochem. Biophys. Res. Comm.* **87**, 734–742.

35 Hoffman, W.L. and O'Shannessy, D.J. (1988) *J. Immunol. Meth.* **112**, 113–120.

36 O'Shannessy, D.J. and Quarles, R.H.

(1985) *J. Appl. Biochem.* **7**, 347–355.

37 O'Shannessy, D.J., Voorstad, P.J. and Quarles, R.H. (1987) *Anal. Biochem.* **163**, 204–209.

38 Heitzmann, H. and Richards, F.M. (1974) *Proc. Natl. Acad. Sci. USA* **71**, 3537–3541.

39 Skutelsky, E. and Bayer, E.A. (1983) *J. Cell Biol.* **96**, 184–190.

40 Roffman, E., Meromsky, L., Ben-Hur, H., Bayer, E.A. and Wilchek, M. (1986) *Biochem. Biophys. Res. Comm.* **136**, 80–85.

41 Bayer, E.A., Ben-Hur, H. and Wilchek, M. (1988) *Anal. Biochem.* **170**, 271–281.

42 Grabarek, Z. and Gergely, J. (1990) *Anal. Biochem.* **185**, 131–135.

43 Rosenberg, M.B., Hawrot, E. and Breakefield, X.O. (1986) *J. Neurochem.* **46**, 641–648.

44 Gilles, M.A., Hudson, A.Q. and Borders, Jr., C.I. (1990) *Anal. Biochem.* **184**, 244–248.

45 Stuchbury, T., Shipton, M., Norris, R. and Malthouse, J.P.G. (1975) *Biochem. J.* **151**, 417–432.

46 Duncan, R.J.S., Weston, P.D. and Wrigglesworth, R. (1983) *Anal. Biochem.* **132**, 68–73.

47 Jue, R., Lambert, J.M., Pierce, L.R. and Traut, R.R. (1978) *Biochemistry* **17**, 5399–5405.

48 Yoshitake, S., Yamada, Y., Ishikawa, E. and Masseyeff, R. (1979) *Eur. J. Biochem.* **101**, 395–399.

49 Crestfield, A.M., Moore, S. and Stein, W.H. (1963) *J. Biol. Chem.* **238**, 622–627.

50 Gurd, F.R.N. (1967) *Methods Enzymol.* **11**, 532–541.

51 Means, G.E. and Feeney, R.E. (1971) In: Protein Modification, Holden-Day, San Francisco, CA, p. 112.

52 Das, M. and Fox, C.F. (1979) *Ann. Rev. Biophys. Bioeng.* **8**, 165–193.

53 Staros, J.V. (1980) TIBS, Dec., 320–322.

54 Forster, A.C., McInnes, J.L., Skingle, D.C. and Symons, R.H. (1985) *Nucl. Acids Res.* **13**, 745–761.

55 Keller, G.H., Huang, D.-P. and Manak, M.M. (1989) *Anal. Biochem.* **177**, 392–395.

Synthesis of Photocleavable Biotinylated Ligands and Application for Affinity Chromatography

Christoph Thiele and
Falk Fahrenholz

Summary

A method for affinity purification based on the high-affinity interaction of biotin-streptavidin and elution by irradiation with UV light was developed. As an example of this approach a biotinylated derivative of the peptide hormone cholecystokinin (CCK-8s) containing a photocleavable *o*-nitrobenzylester group was synthesized. The analog retained high affinity to both CCK receptors and anti-CCK antibodies. It bound to a streptavidin-agarose affinity matrix and was subsequently released by irradiation with UV light of wavelengths >320 nm. For affinity purification of specific antibodies, the modified CCK-8s was incubated with anti-CCK antiserum, and the complex was subsequently passed over a streptavidin-agarose affinity matrix. After washing, the bound antibodies were eluted by photocleavage of the affinity ligand. The eluted antibodies showed essentially unchanged binding characteristics. The approach may prove to be generally useful in the isolation of labile proteins in an intact form.

2.1 Introduction

Affinity chromatography, especially with biotinylated ligands, is a powerful tool for the purification of proteins (1). In contrast to normal affinity chromatography, the use of biotinylated affinity ligands allows protein-ligand binding to take place in solution. The complex formed is then adsorbed to immobilized streptavidin or avidin via the high-affinity biotin-streptavidin (avidin) interaction. To recover the bound protein, the interaction either between protein and ligand or between biotin and avidin has to be disrupted. To break the bond between biotin and avidin or streptavidin, very harsh denaturating conditions (heating with 2% SDS or treatment with 70% formic acid) are required. As no functional protein can be obtained this way, it seems more promising to cleave the complex between ligand and the protein to be purified. But often, especially in the case of antibodies and receptors, this interaction is of high affinity, too, and irreversible denaturation is caused by conditions that disrupt the interaction. To avoid these difficulties, spacers between biotin and the protein-binding part of the ligand are introduced, which can be cleaved by sulfhydryl reagents like mercaptoethanol or DTT (2). However, many proteins that contain disulfide bridges essential for their function, for example, antibodies, cannot be recovered in a functional state by elution with sulfhydryl reagents. Thus, our idea was to use light for elution of material bound to an affinity matrix. In this report we present an efficient method of affinity purification based on the biotin-streptavidin system and elution by irradiation with light as the essential step. The method is demonstrated by the purification of specific antibodies against the peptide hormone cholecystokinin. For this purpose a biotinylated derivate of the C-terminal cholecystokinin octapeptide (CCK-8s) containing a photocleavable *o*-nitrobenzylester group was prepared. The synthetic pathway and affinity chromatography combined with photoelution should be generally useful in isolating labile proteins.

2.2 Technical Procedures

General

Antibodies against CCK peptides were raised in New Zealand rabbits. As an antigen we used benzoylbenzoyl-Orn(propionyl)-CCK-8s (**4**), which had been coupled photochemically by its benzoylbenzoyl group to bovine thyroglobulin. The degree of substitution was 20–50 peptide molecules per thyroglobulin molecule.

Membrane preparation. The preparation of pig brain cortical membranes and the solubilization of membranes with digitonin were performed as described previously (4).

Photolysis. As a source of light we used a high-pressure mercury lamp HBO 200 (Osram), which was built in a housing with a reflector and a focusing lens (E. Leitz, Wetzlar). The sample solution was poured into a quartz cuvette, which was placed in a cooled metal block open to the light source. A glass filter (WG 320, Schott, Mainz) was used to cut off light with wavelengths shorter than 320 nm.

Gel electrophoresis. SDS-polyacrylamide gel electrophoresis was performed according to Laemmli (5) in 7.5 cm × 8.5 cm × 1 mm slab gels. Protein samples were precipitated by methanol/chloroform (6) and were redissolved in sample buffer containing 25 mM DTT before electrophoresis. The gels were stained with silver (7).

Binding assay. Binding studies with digitonin solubilized membranes were performed by a rapid filtration assay as described (8). Briefly, 30 µg of protein from solubilized membranes was incubated with radioactive ligand for 45 min at 20 °C. Then the protein was precipitated by addition of polyethylene glycol and bovine γ globulin to a final concentration of 7% and 0.075%, respectively, and the free ligand was separated by filtration over GF/C glass-fiber filters (Whatman). Nonspecific binding was determined in the presence of 1 µM unlabeled pentagastrin. Binding of radiolabeled peptides to antisera was measured using the same method. But instead of bovine γ globulin, here gelatin (0.2%) was used as a carrier protein for precipitation by polyethylene glycol.

Binding to streptavidin-agarose beads and photoelution. A solution of [^{3}H]BANA-CCK-8s (3.3 nM) in 2 ml of assay buffer [10 mM Hepes/NaOH (pH 7.4), 120 mM NaCl, 5 mM $MgCl_2$, 1 mM EGTA, 0.05% Bacitracin, 0.002% soybean trypsin inhibitor] was mixed with 20 µl of streptavidin-agarose beads. After incubation in the dark for 60 min, the mixture was transferred into a quartz cuvette and was irradiated with light for 2 hours. At all stages of the experiment, aliquots were taken, the streptavidin-agarose was separated by centrifugation, and the content of radioactivity in the solution was determined by scintillation counting.

Affinity chromatography. Anti-CCK antiserum (40 µl) was diluted 100-fold with assay buffer, and [^{3}H]BANA-CCK-8s (10 nM) was added. For determination of nonspecific binding, a second sample containing additionally 10 µM pentagastrin was prepared or preimmune serum was used instead of immune serum. After 3 hours of incubation at 4 °C, streptavidin-agarose (12.5 µl/ml) was added, and the mixture was stirred slowly for 3 hours. Then the agarose beads were separated by centrifugation and were washed with assay buffer (3 × 4 ml). The beads were resuspended in 2 ml of assay buffer and were irradiated at wavelengths >320 nm for 40 min at 4 °C. During irradiation the suspension was stirred magnetically to prevent sedimentation of the agarose beads.

Materials

All amino acid derivatives and peptides were from Bachem, Heidelberg. *p*-Toluic acid was purchased from Aldrich, digitonin (water soluble type), biotin-ε-Ahx and streptavidin-agarose from Sigma. [^{3}H]*N*-succinimidyl propionate (specific activity 100 Ci/mmol, 37.5 TBq/mmol) was from Amersham. The materials for gel electrophoresis, including the molecular weight standard proteins, were from Bio-Rad.

Peptides were separated by reversed-phase high-performance liquid chromatography (HPLC). All separations were carried out on a Varian 5000 liquid chromatograph using a 10-µm LiChrosorb RP 18 column (24 × 0.46 cm). The gradient system used was buffer A: 25 mM triethylammonium phosphate pH 3.4 in water; buffer B: 10% buffer A + 90% acetonitrile. Gradient: 0–5 min constantly A/B 75/25, 5–40 min 75/25 to 0/100 with UV detection at 220 nm. Protein determination was performed by the fluorescamine method using a Jobin Yvon JY 3-D spectrofluorometer. Bovine serum albumin was used as a standard protein.

Chemical Synthesis

Procedure 2.1

Caesium salt of biotin-aminohexanoic acid (biotin-ε-Ahx-Cs^+) (1)

A total of 100 mg (0.28 mmol) of biotin-ε-Ahx was dissolved in 2 ml of water containing 45.5 mg (0.14 mmol) of Cs_2CO_3. The solvent was removed by evaporation under reduced pressure. Toluene (0.5 ml) was added, and again all liquid was evaporated. This procedure was repeated 3 times to give the anhydrous cesium salt with quantitative yield.

Procedure 2.2

3-Nitro-4-bromomethylbenzoic acid 2′,4′,5′ trichlorophenylester (2)

A total of 0.50 g (2.05 mmol) of 3-nitro-4-bromomethylbenzoic acid (3) and 0.40 g (2.03 mmol) of 2,4,5-trichlorophenol was dissolved in 40 ml of ethyl acetate. To this mixture was added 0.42 g (2.04 mmol) of *N,N′*-dicyclohexyl carbodiimide (DCC) in 10 ml of ethyl acetate. After stirring for 16 hours, the white precipitate was removed by filtration, and the solvent was evaporated. The residue was recrystallized from CH_2Cl_2/pentane to give a slightly yellow product. Yield: 700 mg (1.65 mmol, 81%). Melting point 124–125 °C. HPLC: Retention time (R_t) = 39.0 min. ^{1}H-NMR ($CDCl_3$): δ 8.83 (d, *J* = 1.8 Hz, 1H, H-2), 8.40 (dd, *J* = 1.8, 8.2 Hz, 1H, H-6), 7.79 (d, *J* = 8.2 Hz, 1H, H-5), 7.63 (s, 1H, H-3′), 7.45 (s, 1H, H-6′), 4.88 (s, 2H, benzyl-H).
Positive ion electron ionization mass spectroscopy (EI-MS): *m/z* = 437 $[M + H]^+$.

Procedure 2.3

3-Nitro-4-(biotin-ε-Ahx-oxymethyl)benzoic acid 2′,4′,5′-trichlorophenylester (3)

A total of 95 mg (224 µmol) of 3-nitro-4-bromomethylbenzoic acid 2′,4′,5′-trichlorophenylester was dissolved in 7.5 ml of *N*-methylpyrrolidone. This solution was chilled to 4 °C and was added to a suspension of 100 mg (204 µmol) of biotin-ε-Ahx-Cs^+ in 7.5 ml of *N*-methylpyrrolidone, and the mixture was stirred for 16 hours at 4 °C. The clear solution was diluted with 60 ml of ethyl acetate, and the organic phase was subsequently washed with carbonate buffer (pH 8.8, 3 × 20 ml) and water (3 × 20 ml). The solvent was partially evaporated, and the product was precipitated with ether to give 110 mg (154 μmol, 75%) of product of approximately 80% purity, which was used for further synthesis without prior purification. An analytical sample was purified by semipreparative HPLC. HPLC: (R_t) = 32.4 min. Positive ion (FAB-MS): m/z = 715 $[M + H]^+$, 536 $[M - C_6H_2Cl_3 + H]^+$.

Procedure 2.4

4-(Biotin-ε-Ahx-oxymethyl)-3-nitrobenzoyl-Gly-Orn(Fmoc)-OH (4)

A total of 20 mg (38 μmol) of H-Gly-Orn(Fmoc)-OH was dissolved in 0.5 ml of 1.6% Na_2CO_3 in water and 0.5 ml of 1,4-dioxan. Twenty-seven milligrams (38 μmol) of 3-nitro-4-(biotin-ε-Ahx-oxymethyl)benzoic acid 2′,4′,5′-trichlorophenylester dissolved in 100 μl of dimethylformamide (DMF) was added, and the mixture was allowed to stand for 2 hours. Ether (10 ml) was added, the upper phase was discarded, and the product was precipitated by the addition of 1% citric acid in water. Further purification was achieved by semipreparative HPLC. HPLC: R_t = 23.4 min. Electrospray MS: m/z = 930.5 $[M + H]^+$.

Procedure 2.5

ε-Ahx-CCK-8s (5)

A total of 37 mg (32 μmol) of desulfated cholecystokinin octapeptide trifluoroacetate (CCK-8ds × TFA) [prepared from Boc-CCK-8ds by deprotection with trifluoroacetic acid (TFA)] was dissolved in 500 ml of dimethylformamide (DMF) and 17.3 ml (130 μmol) of triethylamine. A solution of 36 mg (80 μmol) of Fmoc-ε-Ahx-ONSu in 500 μl of DMF was added. After 2 hours the peptide was precipitated by addition of 15 ml of ice-cold ether/ethyl acetate 2/1.

The precipitate was dissolved in 400 μl of DMF/pyridine 1/1. SO_3-pyridine complex (100 mg, Fluka) was added, and the mixture was stirred for 16 hours. Then excess SO_3-pyridine complex was destroyed with 0.1 M Na_2CO_3 buffer, pH 8.9. The sulfated product was isolated by solid-phase extraction on a Sep-Pak RP 18 cartridge (Millipore) as described (4).

To remove the Fmoc protecting group, the lyophilized product was treated with 1 ml of 20% piperidine in DMF. The final product was purified on HPLC. HPLC: R_t = 16.0 min. Negative ion FAB-MS: *m/z* = 1254 $[M - H]^-$.

Procedure 2.6

4-(Biotin-ε-Ahx-oxymethyl)-3-nitrobenzoyl-Gly-Orn-ε-Ahx-CCK-8s (6)

4-(Biotin-ε–Ahx-oxymethyl)-3-nitrobenzoyl-Gly-Orn(Fmoc)-OH (1.2 mg, 1.3 µmol) was dissolved in 30 µl of DMF. Equimolar amounts of *N*-hydroxysuccinimide and DCC dissolved in DMF were added, and the mixture was allowed to stand for 16 hours. Then a solution of 1.6 mg (1.3 µmol) of ε-Ahx-CCK-8s in 10 µl of DMF and 0.2 µl of triethylamine was added. After 30 min, when more than 90% of the educts had disappeared, 10 µl of piperidine was added to remove the Fmoc protecting group. The product was purified by semi-preparative HPLC. HPLC: R_t = 17.5 min. Electrospray MS: expected *m/z* = 1945.5 $[M + H]^+$, found *m/z* = 1967.5 $[M + Na]^+$, 1983.5 $[M + K]^+$.

Procedure 2.7

[3H]4-(Biotin-ε–Ahx-oxymethyl)-3-nitrobenzoyl-Gly-Orn(propionyl)-ε–Ahx-CCK-8s (7) ([3H]BANA-CCK-8s, see Fig. 2.1)

4-(Biotin-ε–Ahx-oxymethyl)-3-nitrobenzoyl-Gly-Orn-ε–Ahx-CCK-8s (0.21 mg, 0.11 µmol) was dissolved in 30 µl of 0.2% triethylamine in DMF. A solution of 750 mCi (7.5 nmol, specific activity 100 Ci/mmol) of $[^3H]$*N*-succinimidyl propionate in toluene was placed in a gentle stream of nitrogen until the solvent was evaporated. Care was taken to stop the nitrogen stream immediately after evaporation was finished. A total of 3 µl of the above peptide solution was added, and the mixture was allowed to stand at room temperature for 3 hours. Then the pure product was isolated with a radiochemical yield of 33% using HPLC. HPLC: R_t = 18.4 min.

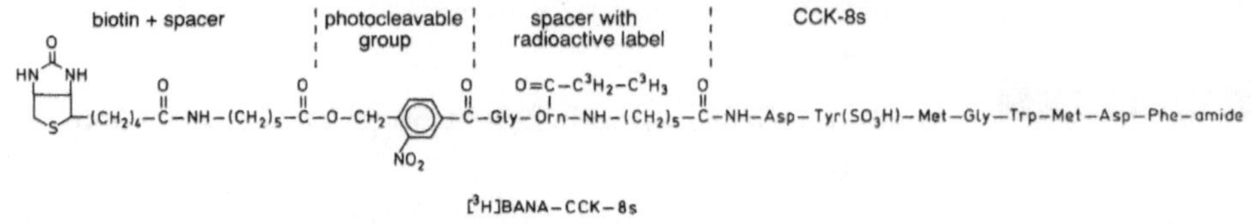

biotin + spacer | photocleavable group | spacer with radioactive label | CCK-8s

$[^3H]$BANA-CCK-8s

Figure 2.1 Formula of the biotinylated photocleavable cholecystokinin analog [^{3}H]BANA-CCK-8s

2.3 Results and Discussion

Chemical synthesis

A biotinylated hormone analog of the CCK-octapeptide (H-Asp-Tyr(SO_3H)-Met-Gly-Trp-Met-Asp-Phe-amide) containing the light-sensitive *o*-nitrobenzyl ester group was synthesized. At the N-terminus of the ligand, biotin was introduced for binding to streptavidin. These 2 parts of the molecule were connected by a light-sensitive *o*-nitrobenzyl ester group, which was originally introduced as a protecting group by Patchornik et al. (9) and used for cleavable resins in synthesis by Rich and Gurwara (10). This ester is stable against acid, base, and reductive agents and can be cleaved between the carboxylic oxygen and the benzylic carbon by light of 350 nm. To facilitate analysis of the binding and the chromatographic steps, the biotinylated peptide moiety was radioactively labeled with the [2,3-^{3}H]propionyl group. To avoid steric hindrance in the ternary complexes streptavidin-ligand-protein, 2 additional ε-aminohexanoic acid groups were introduced as spacers (Fig. 2.1).

The photocleavable, biotinylated, radioactive CCK derivative [^{3}H]BANA-CCK-8s was synthesized by fragment condensation in liquid phase (Fig. 2.2). A crucial step in the synthesis was the first reaction in the scheme, the formation of the light-sensitive nitrobenzylic ester of biotin-ε-aminohexanoic acid. While Fmoc-protected alanine could be coupled to 4-hydroxymethyl-3-nitrobenzoic acid 2',4',5'-trichlorophenylester with DCC and 4-dimethylamino-pyridine (11), this reaction failed if the protected α-amino acid was replaced by biotin-ε-aminohexanoic acid. Thus we used the cesium salt of biotin-ε-aminohexanoic acid (**1**) for direct coupling to the 4-bromomethyl-3-nitrobenzoic acid 2',4',5'-trichlorophenylester (**2**) according to the method described by Hemmasi et al. (12) for the synthesis of 4-(Bocaminoacyloxymethyl)-3-nitrobenzoic acids. The 2,4,5-trichlorophenylester group of **2** served as a protecting group during the nucleophilic substitution reaction with **1**, and activated the product **3** for the following coupling to the α-amino group of the dipeptide H-Gly-Orn(Fmoc)-OH. The ornithine residue was incorporated to label the ligand radioactively at the δ-amino group. Though the glycine residue does not play a role for the function of the peptide, it was of importance for the synthesis, as the following activation/coupling step failed if the substituted nitrobenzoic acid **3** was coupled to ornithine without a glycine residue as a spacer. The CCK moiety (**5**) of the peptide was introduced as a derivative of CCK-8s which had been elongated N-terminally with ε-aminohexanoic acid serving as an additional spacer group. After deprotection of the δ-amino group of ornithine, radioactivity was introduced by acylation with commercially available [2,3-^{3}H]*N*-succinimidyl propionate (Fig. 2.2).

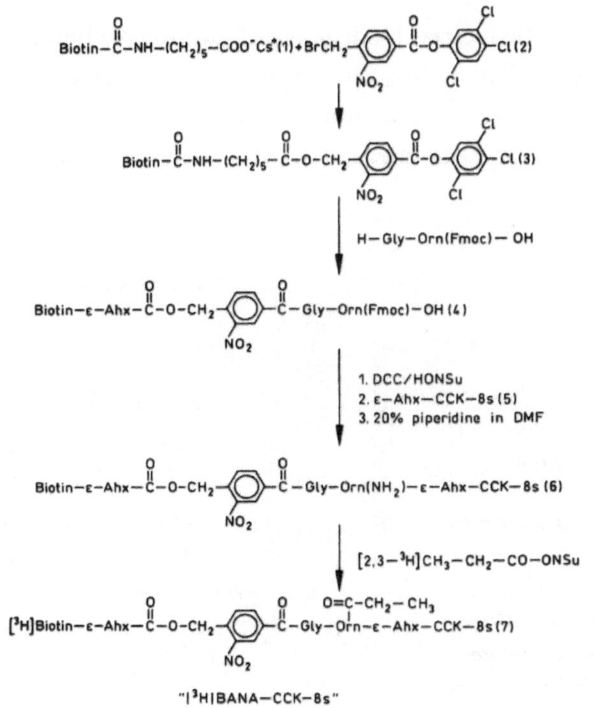

Figure 2.2 Synthesis of [^{3}H]BANA-CCK-8s (7)

Binding to anti-CCK antisera and CCK-B receptors

Binding of [^{3}H]BANA-CCK-8s to anti-CCK antisera occurred in a specific, saturable manner with high affinity. Scatchard analysis of the binding data reveals the existence of a homogeneous population of binding sites ($K_d = 1.5 \pm 0.3$ nM). It also bound with high affinity ($K_d = 0.8 \pm 0.1$ nM) and low nonspecific binding to digitonin-solubilized pig brain cortical membranes.

To examine detrimental effects of photoelution toward proteins, we determined the binding capacity of CCK-B receptor preparations that had been irradiated under the same conditions as used for photoelution. The receptor was chosen for this experiment, as it is more sensitive to denaturation. The maximum binding capacity was reduced by 15%, while the affinity remained unchanged.

Binding to streptavidin agarose and photoelution

The biotinylated CCK derivative [^{3}H]BANA-CCK-8s bound rapidly and nearly completely to streptavidin immobilized on agarose beads (Fig. 2.3). After 30 min, 96% of the ligand was bound to the matrix. No leakage of radioactive ligand could be seen on prolonged incubation. The radioactive peptide was then released from the matrix by irradiation with light of wavelengths >320 nm. Using a 200-W mercury lamp, 53% of the radioactivity was released after 15 min of irradiation and 79% after 120 min.

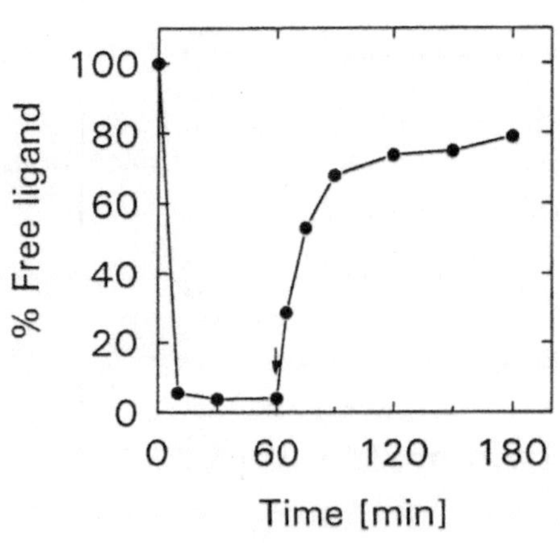

Figure 2.3 Binding of [^{3}H]BANA-CCK-8s to streptavidin-agarose and subsequent photoelution

Two milliliters of a solution of [^{3}H]BANA-CCK-8s (3.3 nM) were incubated with 20 ml of streptavidin-agarose. After 1 hour (as marked by the arrow) irradiation with light of wavelengths >320 nm was started. Aliquots were taken at different times, and the content of free ligand was determined by scintillation counting.

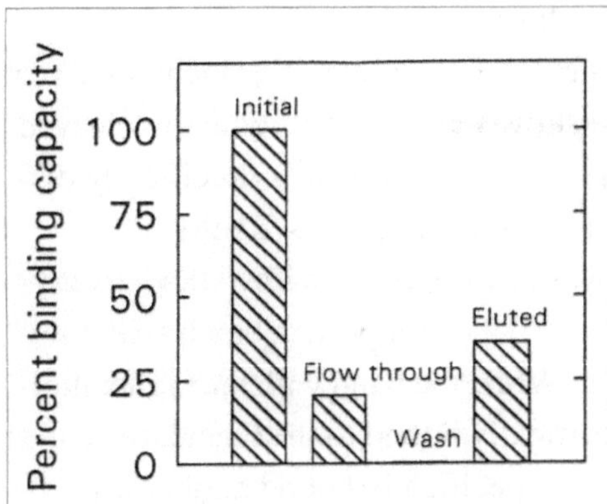

Figure 2.4 Affinity purification of anti-CCK antibodies using [³H]BANA-CCK-8s and streptavidin-agarose

Diluted anti-CCK antiserum was incubated with [³H]BANA-CCK-8s and subsequently mixed with streptavidin-agarose. After washing, bound protein was eluted by irradiation with light of wavelengths >320 nm. In the diagram the relative binding activity in the solution before and after incubation with streptavidin-agarose, in the wash solution, and in the eluted protein fraction is shown.

Affinity chromatography

The usefulness of the concept of photoelution is demonstrated by purification of anti-CCK antibodies. Purification was performed in 3 sequential steps. First, antiserum was incubated with [^{3}H]BANA-CCK-8s. The amount of nonspecific binding was determined by a parallel experiment where 10 μM pentagastrin was added at this stage. Alternatively, preimmune serum was used to distinguish specific from nonspecific binding. Then streptavidin-agarose beads were added, and the suspension was equilibrated for 3 hours at 4 °C to allow complete binding of the excess of biotinylated ligand and of the ligand-antibody complex to the matrix. Ninety percent of the ligand and 79% of anti-CCK antibodies (Fig. 2.4) were adsorbed on the immobilized streptavidin in this step. The unbound proteins were washed away, and the CCK-binding proteins were specifically eluted by irradiation with light. The amount of eluted specific binding activity after 40 min of irradiation was 36% of the initial value or 46% of that bound to the matrix, respectively. The total amount of protein eluted from the matrix was about 0.1% of the starting material. The calculated enrichment was 360-fold. To demonstrate the specificity of the eluted binding activity, the eluate was mixed with increasing concentrations of pentagastrin, which shares the epitope Trp-Met-Asp-Phe-amide with [^{3}H]BANA-CCK-8s, and after incubation for 16 hours binding was measured. The results show that the photolytically cleaved [^{3}H]BANA-CCK-8s, which was bound to the eluted proteins, was specifically displaced by pentagastrin. The calculated inhibition constant K_i was 20 nM.

The composition of the purified material was also analyzed with SDS-polyacrylamide gel electrophoresis. As shown in Figure 2.5, 2 major protein bands with molecular weights of 95 kD and 52 kD were eluted. While the band with the higher molecular weight was caused by proteins nonspecifically bound to the ligand, the band with 52 kD corresponded to the heavy chains of specifically bound antibodies.

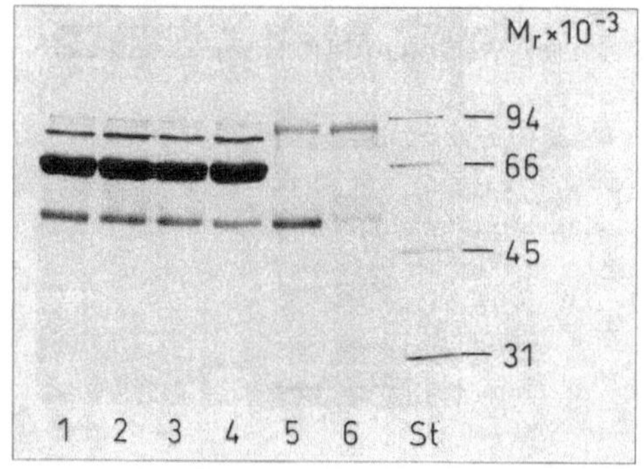

Figure 2.5 SDS-PAGE analysis of the purification of anti-CCK antibodies from antiserum using the [^{3}H]BANA-CCK-8s/streptavidin system

Lanes 1, 3, 5: anti-CCK antiserum; lanes 2, 4, 6: preimmune serum; St: standard proteins. Lanes 1 and 2 show the probe before incubation with streptavidin-agarose, lanes 3 and 4 after incubation with streptavidin-agarose, lanes 5 and 6 photoeluted proteins. The specifically eluted band with M_r of 52 kD in lane 5 represents the heavy chains of the specific anti-CCK antibodies. The protein content of lanes 1 to 4 corresponds to 0.1% of the applied protein, while lanes 5 and 6 contain 33% of the eluted proteins.

The concept of photoelution demonstrated here has considerable advantages compared with other elution techniques (13):

1) The eluting "agent," UV light of long wavelength (>320 nm), does not usually interfere with the biological activity of proteins or even living cells (14).
2) Elution is highly specific because only bonds between the matrix and the ligand are cleaved. Substances which are nonspecifically adsorbed to other components of the matrix should be retained on the matrix.
3) The composition of the buffer used during elution can be selected to provide the optimum conditions for the stability of eluted proteins. No buffer change or dialysis has to be performed after photoelution.

The method of photoelution presented here is not limited to the purification of CCK-binding proteins. Both biotinylated compounds **3** and **4** can be coupled to any peptide or other ligand containing a free amino group, and if compound **4** is used, radioactivity or fluorescent reporter groups can be introduced additionally into the molecule. The activated ester **3** would also be a suitable reagent for the immediate use in solid-phase peptide synthesis. A possible application of the approach is the purification of other antibodies, but any other protein which binds with a sufficiently high affinity to a suitable ligand might also be purified.

2.4 Troubleshooting

The analytical scale method described here might also be scaled up for preparative purposes. The lamp we used would allow irradiation of samples of up to 2 ml of (streptavidin-)affinity matrix, while even larger amounts could be irradiated in a photoreactor like the Rayonet RPR 100 (Southern New England Ultraviolet Company, Hamden) equipped with 350-nm lamps.

High concentrations of substances strongly absorbing in the wavelength range from 340 to 360 nm should be omitted during photoelution because their presence would result in prolonged irradiation times.

References

1 Wilchek, M. and Bayer, E.A. (1990) *Methods in Enzymology*. Vol. 184, Academic Press, San Diego, CA.

2 Shimkus, M., Levy, L. and Herman, T. (1985) *Proc. Natl. Acad. Sci. USA* **82**, 2593–2597.

3 Barany, G. and Albericio, F. (1985) *J. Am. Chem. Soc.* **107**, 4936–4942.

4 Thiele, C. and Fahrenholz, F. (1993) *Biochemistry* **32**, 2741–2746.

5 Laemmli, U.K. (1970) *Nature* **227**, 680–685.

6 Wessel, D. and Flügge, U.I. (1984) *Anal. Biochem.* **138**, 141–143.

7 Wray, W., Boulakis, T., Wray, V.P. and Hancock, R. (1981) *Anal. Biochem.* **118**, 197–203.

8 Gut, S.H., Demouliou-Mason, C.D., Hunter, J.C., Hughes, J. and Barnard, E.R. (1989) *Eur. J. Pharmacol.* **172**, 339–346.

9 Patchornik, A., Amit, B., and Woodword, R.B. (1970) *J. Am. Chem. Soc.* **92**, 6333.

10 Rich, D.H. and Gurwara, S.K. (1975) *J. Am. Chem. Soc.* **97**, 1575–1579.

11 Kneib-Cordonier, N., Albericio, F. and Barany, G. (1990) *Int. J. Peptide Protein Res.* **35**, 527–538.

12 Hemmasi, B., Stueber, W. and Bayer, E. (1982) *Physiol. Chem. Hoppe-Seyler* **363**, 701–708.

13 Thiele, C. and Fahrenholz, F. (1994) *Anal. Biochem.* **218**, 330–337.

14 McLaren, A.D. and Shugar, D. (1964) *Photochemistry of Proteins and Nucleic Acids*. Pergamon Press, Oxford.

Purification of the Receptor for Pituitary Adenylate Cyclase-Activating Polypeptide (PACAP) using Biotinylated Ligands

Tetsuya Ohtaki, Chieko Kitada and Haruo Onda

Summary

The pituitary adenylate cyclase-activating polypeptide (PACAP), including PACAP27 and PACAP38, stimulates adenylate cyclase via specific PACAP receptors that are coupled to G proteins. The present study describes the application of avidin-biotin technology to the purification of the PACAP receptor from bovine brain membranes. The PACAP receptor was solubilized from the membranes with digitonin and partially purified by DEAE-Toyopearl and hydroxylapatite chromatography for removal of avidin-binding proteins. The partially purified PACAP receptor was mixed with a biotinylated ligand (PACAP27-Cys(biotin)-NH_2) to form a receptor-ligand complex, adsorbed onto avidin-agarose, and then eluted with 1.0 M NaCl-acetate buffer (pH 4.0). The affinity-purified receptor was further purified by hydroxylapatite and gel-filtration chromatography. Analysis of the purified preparation by SDS-polyacrylamide gel electrophoresis and silver staining showed a single broad protein band with an M_r = 55 000 to 60 000. The specific activity of the purified receptor (17.2 nmol/mg) was close to the theoretical value. The dissociation constant (K_d = 25.8 pM) was similar to that of the crude solubilized PACAP receptor. Thus, the PACAP receptor was purified to near homogeneity in a fully active form preserving high affinity for PACAP. The present study demonstrates that affinity chromatography using biotinylated ligands is a promising method for the purification of the PACAP receptor.

3.1 Introduction

The high-affinity interaction between avidin and biotin was first employed in receptor purification by Hofmann and his colleagues in 1984 (1). They developed a novel affinity chromatography method using a biotinylated ligand and immobilized avidin for purification of the insulin receptor (1). Similar applications to purification of various receptors such as the estrogen receptor (2) and the GnRH receptor (3) were also described in 1985–1986.

As detailed in an excellent review (4), this strategy requires a biotinylated ligand that is able to bind simultaneously to its receptor and avidin with high affinity. To meet this criterion, (i) a ligand should be derivatized at a specific position that is minimally involved in receptor binding, and (ii) a biotin moiety should be separated from a ligand via a long spacer arm. Finn and Hofmann demonstrated that biotinylated insulin derivatives with long spacer arms dissociated from succinylated avidin at a slower rate than those with short spacer arms. A long spacer arm is presumably required to decrease steric hindrance between insulin and avidin. Indeed, it is well known that the biotin-binding site is buried in deep clefts (about 10 Å depth) in the avidin structure. Moreover, a long spacer arm diminishes possible steric hindrance between avidin and the receptor on the avidin-biotinylated ligand-receptor complex. Redeuilh et al. reported that the presence of avidin in the binding reaction mixture largely decreased the affinity of short arm-biotinylated estradiols for their receptors but had less effect on long arm-biotinylated derivatives (5).

In the field of G protein-coupled receptors, several applications of avidin-biotin technology to receptor purification have been described recently, including purification of the endothelin receptor (6), the somatostatin receptor (7, 8), and the pituitary adenylate cyclase-activating polypeptide (PACAP) receptor (9, 10). These followed the previous application (3) to the GnRH receptor (a G protein-coupled receptor) principally; however, several modifications were required to solve problems in each application. For the somatostatin receptor, a substantial decrease in binding affinity during solubilization, a problem often encountered in purification of G protein-coupled receptors, obstructed straightforward application. This problem was successfully bypassed using a modified procedure: (i) solubilization of the preformed stable complex of biotinylated ligand-receptor-G protein; (ii) retrieval of the solubilized complex by immobilized avidin; and (iii) elution of receptor by treatment with GTP (or GDP + AlF_4^-), which dissociates G proteins and converts the receptor into a low-affinity form (7, 8). This was an excellent application demonstrating the advantage of affinity chromatography using biotinylated ligands in this field.

For our purification of the PACAP receptor, a similar problem in solubilization was solved by the use of digitonin (9). An avidin-binding protein found in the membrane solubilizate that interfered with this application seriously was removed by conventional chromatography methods. The use of digitonin in this partial purification was advantageous in isolating the PACAP receptor in a high affinity form. The combination of these techniques and affinity chromatography method using biotinylated PACAP led to the successful purification of the PACAP receptor near to homogeneity. The following section describes the entire procedure for purifying of the PACAP receptor (Fig. 3.1), including solubilization and partial purification methods as well as the application of the avidin-biotin high-affinity interaction.

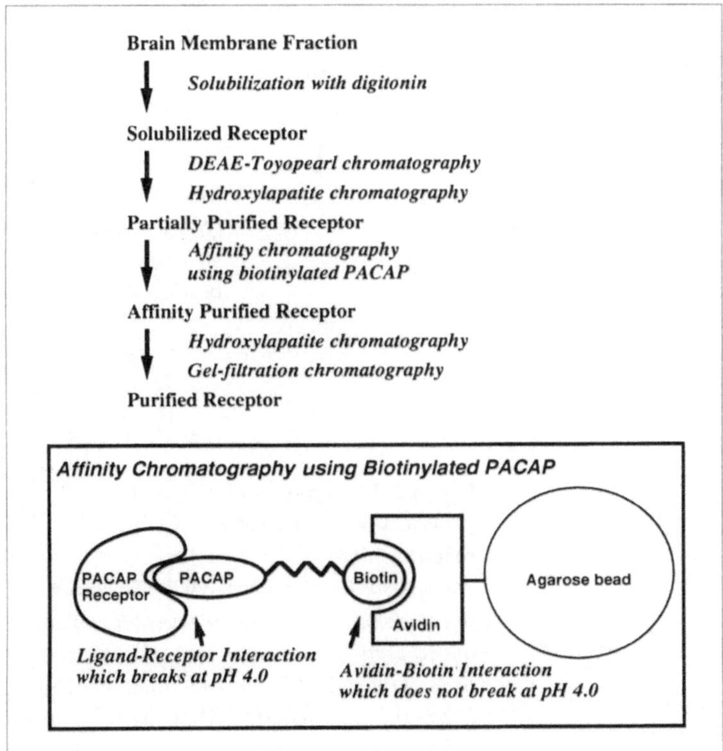

Figure 3.1 *The procedure for purification of the PACAP receptor from bovine brain membranes*

3.2 Technical Procedures

Preparation of a Biotinylated Ligand

The PACAP27 ligand has several potential biotinylation sites, including 1 α-amino and 3 ε-amino residues. However, it is difficult to restrict the reaction to a specific desired site. To get around this problem, we introduced a Cys^{28} residue at the carboxy terminus of the PACAP27 peptide and biotinylated to its thiol group using a biotinylating reagent *N*-[6-(biotinamido)hexyl]-3'-(2'-pyridyldithio)propionamide (biotin-HPDP). The reagent has a long spacer arm (29.2 Å) between its biotin moiety and pyridyldithio group (Fig. 3.2). As shown in Figure 3.3, the biotinylated ligand PACAP27-Cys(biotin)-NH_2 (the structure is shown in Fig. 3.2) displaced the binding of the radiolabeled PACAP27 potently both in the presence and absence of avidin. This result indicates that the biotinylated PACAP can bind to the PACAP receptor and avidin to form a stable ternary complex.

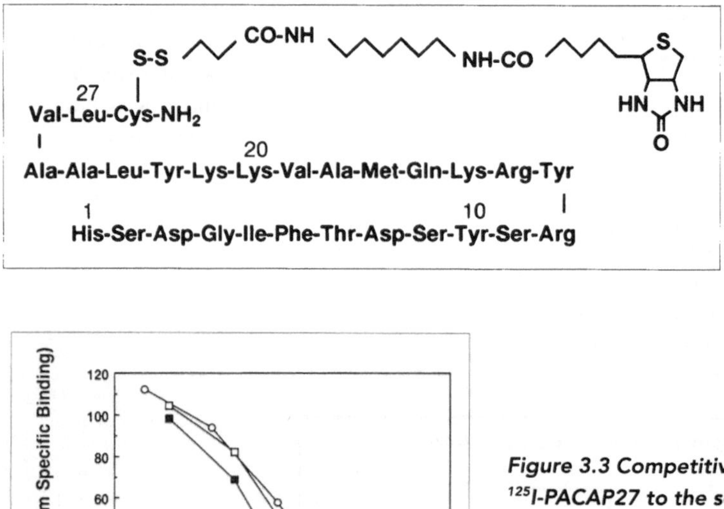

S-S CO-NH NH-CO S

27
Val-Leu-Cys-NH_2 HN NH O

20
Ala-Ala-Leu-Tyr-Lys-Lys-Val-Ala-Met-Gln-Lys-Arg-Tyr

1 10
His-Ser-Asp-Gly-Ile-Phe-Thr-Asp-Ser-Tyr-Ser-Arg

Figure 3.2 The structure of the biotinylated PACAP

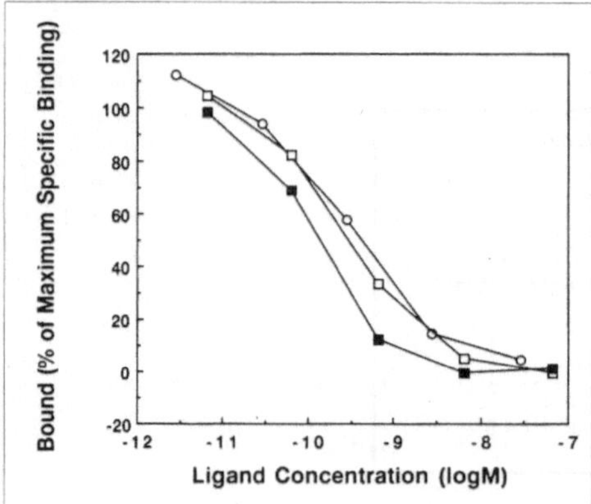

Figure 3.3 Competitive binding of ^{125}I-PACAP27 to the solubilized PACAP receptor

The specific binding of ^{125}I-PACAP27 (100 pM) to the crude solubilized PACAP receptor was determined in the presence of increasing concentrations of PACAP27 (○), the biotinylated PACAP27 (□), or the biotinylated PACAP27-avidin complex (final avidin concentration was 0.3 µM) (■).

Protocol 3.1

Preparation of PACAP27-Cys(biotin)-NH_2

1. Dissolve 10 mg (about 3 μ mol) of lyophilized PACAP27-Cys-NH_2 powder in 18 ml of ice-cold 10 mM potassium phosphate buffer (pH 7.0) containing 3 mM EDTA and 30% acetonitrile.
2. Dissolve 18 mg (10 molar equivalents) of biotin-HPDP in 6 ml of dimethyl sulfoxide (DMSO).
3. Add the biotin-HPDP solution to the chilled peptide solution drop by drop. Incubate the reaction mixture at 4 °C overnight.
4. Dilute the reaction mixture with 16 ml of 0.05% trifluoroacetic acid (TFA).
5. Inject the diluted mixture (10 ml aliquot for each run) into an ODS80TM high-performance liquid chromatograph (HPLC) column (7.6 mm × 30 cm, Tosoh Corporation, Tokyo) equilibrated with 0.05% TFA. Elute the biotinylated PACAP27 with a linear-gradient increase of acetonitrile concentration from 20% to 40% in 60 min at a flow rate of 2.5 ml/min (Fig. 3.4).
6. Collect the peak of the biotinylated PACAP27 and subject it to rechromatography under the same conditions.
7. Combine and lyophilize the purified peptide in a SpeedVac concentrator (Savant) without heating.
8. Dissolve the peptide in 25 ml of distilled water and determine the peptide concentration by the method of Murphy and Kies (11).
9. Immediately add 1/30 volume of 3% 3-[(3 cholamidopropyl)dimethylammonio]propanesulfonic acid (CHAPS) solution to the peptide solution. Store the peptide at –70 °C. It should be noted that the peptide tends to adhere strongly to the tube wall in the absence of CHAPS. This procedure usually provides 0.6 μmol of the biotinylated PACAP (about 25 μM peptide solution).

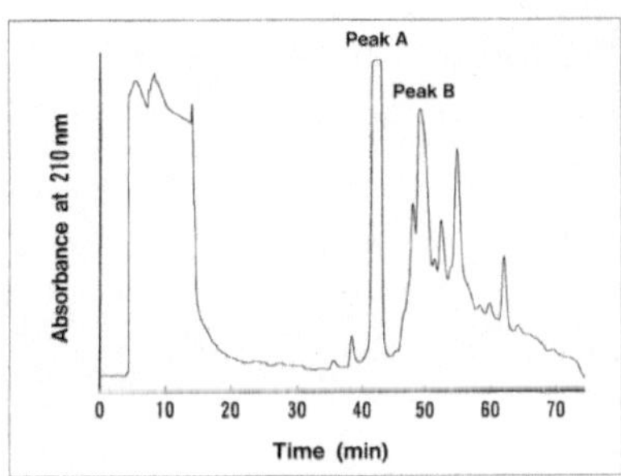

Figure 3.4 Purification of the biotinylated PACAP by reversed-phase HPLC

The reaction mixture was separated by reversed-phase HPLC as described in Technical Procedures. Peak A is Biotin-HPDP, Peak B is the biotinylated PACAP27. Peak B was rechromatographed to obtain pure preparation of the biotinylated PACAP27.

Preparation of Membrane Fraction

We usually prepare the membrane fraction from 10 bovine brains at once. The 1-day procedure for this large-scale preparation requires a zonal-type ultracentrifuge rotor (the rotor employed in this study was an RPZ35T rotor acquired from Hitachi). Alternatively, this procedure could be divided into several batches. Membranes may be kept frozen at –70 °C for as long as one year without significant loss of activity. All of the following purification procedures should be carried out at 4 °C.

Protocol 3.2

Preparation of membrane fraction from bovine brains

1. Remove the brain stem, cerebellum, meninges, and blood clots from fresh bovine brains.
2. Perform a rough homogenization of the dissected tissue (3 kg from 10 brains) in 4 volumes of 0.25 M sucrose-TED_{10} buffer (20 mM Tris, 10 mM EDTA, 0.25 M sucrose, 0.03% NaN_3, 0.5 mM PMSF, 1 μg/ml pepstatin, 20 μg/ml leupeptin, 4 μ g/ml E-64; pH 7.4) using a Waring blender.
   Follow with a thorough homogenization using a Polytron homogenizer (Kinematica GmbH).
3. Centrifuge the homogenate at 10 000 × g for 15 min to obtain the supernatant.
4. Ultracentrifuge the supernatant at 100 000 × g for 1 hour.
5. Suspend the resulting pellet in 0.15 M NaCl-TED_5 buffer (20 mM Tris, 5 mM EDTA, 0.03% NaN_3, and the protease inhibitors listed above; pH 7.4). Ultracentrifuge the suspension again.
6. Repeat step 5, once with 0.15 M NaCl-TED_5 buffer and once with TED_5 buffer.
7. Suspend the membrane pellet in TED_5 buffer at 7 to 10 mg protein/ml. Store it at –70 °C.

Solubilization of the PACAP Receptor

Solubilization of receptor in a high-affinity form is prerequisite to its affinity purification. Hence, the first step of receptor purification is a screening of detergents. As shown in Figure 3.5, most detergents at higher concentrations easily denature the solubilized receptor, and thus require careful monitoring during all purification procedures. However, we found that even an excess of digitonin (a nonionic detergent) failed to eliminate PACAP receptor activity (Fig. 3.5), so we used digitonin in the present study.

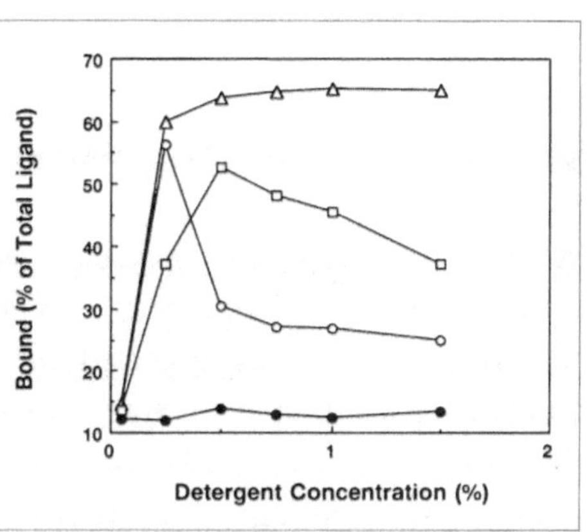

Figure 3.5 Solubilization of the PACAP receptor with various detergents
The brain membranes (1 mg protein/ml) were solubilized with the indicated concentrations of CHAPS (○), digitonin (△), BIGCHAP (□), or octylglucoside (●). Total ^{125}I-PACAP27 binding (including nonspecific binding) was determined with the solubilized materials diluted 10-fold with 0.05% digitonin-0.1% BSA-TED buffer.

The use of digitonin does pose several technical difficulties. One problem is its heterogeneity and solubility in water. To prepare a 5% (w/v) solution, the detergent must first be boiled for 30 min in the appropriate volume of water to dissolve digitonin and then left at room temperature for 48 hours. If insoluble parts remain, the solution is then filtered through a 0.22 μm cartridge filter and the filtrate is lyophilized. The resulting digitonin preparation is then completely water soluble. Alternatively, a digitonin preparation that is soluble in water at room temperature is available from Wako (Osaka, Japan). Another problem is its low critical micelle concentration and large micelle size, which make it difficult to remove by dialysis. Removal of digitonin from the receptor preparation is done with hydroxylapatite chromatography (see Protocol 3.6).

Protocol 3.3 **Solubilization of the PACAP receptor with digitonin**

1. Thaw the membranes.
2. Prepare a mixture of the membrane suspension and digitonin-TED_5 buffer that contains 1.5% digitonin and 3 mg protein/ml (digitonin:protein = 5:1).
3. Stir the mixture for 1 hour using a magnetic stirrer.
4. Ultracentrifuge the mixture at 100 000 × g for 1 hour to obtain the supernatant.
5. Subject the supernatant to partial purification, or store it frozen at –70 °C.

Partial Purification of the PACAP Receptor

We found that an intermediate step was necessary before affinity chromatography of the solubilized crude proteins. When the solubilized material from the brain membrane fraction was subjected directly to biotinylated PACAP-avidin-agarose affinity chromatography, a large amount of a protein with an M_r = 57 000 (named 57K protein) was co-eluted with the ligand (^{125}I-PACAP27)-binding activity in the acidic eluate. Although the molecular weight of the 57K protein was compatible with that of the receptor predicted from the affinity-labeling experiments (12), it was, unfortunately, not the receptor. The 57K protein was found to be an avidin-binding protein via a negative control experiment done in the absence of biotinylated ligand.

The binding mechanism of the 57K protein to avidin is not clear. Nonspecific adsorption of acidic proteins to avidin (a basic protein) has been reported, and can be decreased by the use of succinylated avidin. In the present study, we conducted an intermediate partial-purification step to remove the 57K protein using a combination of DEAE ion-exchange and hydroxylapatite chromatography. A reduction in the volume of the solubilizate is another advantage of this procedure.

For these chromatographic steps, the choice of detergent is important to isolate PACAP receptors in an active form. DEAE ion-exchange chromatography in the presence of CHAPS did not yield significant PACAP receptor activity. While *N,N*-bis(3-D-gluconamidopropyl) cholamide (BIGCHAP) and octylglucoside provided better results, the best result was obtained with digitonin (Fig. 3.6). Hydroxylapatite chromatography in the presence of digitonin was effective for purification because an unusually high concentration of potassium buffer (higher than 0.4 M) was required for elution of the PACAP receptor, whilst most proteins were eluted with 0.2 to 0.3 M potassium phosphate buffer. Chromatography in the presence of CHAPS was not successful, because the PACAP receptor was largely denatured and co-eluted in earlier fractions with the bulk of the proteins. Thus, digitonin proved to be the most appropriate detergent for partial purification of the PACAP receptor by conventional chromatography.

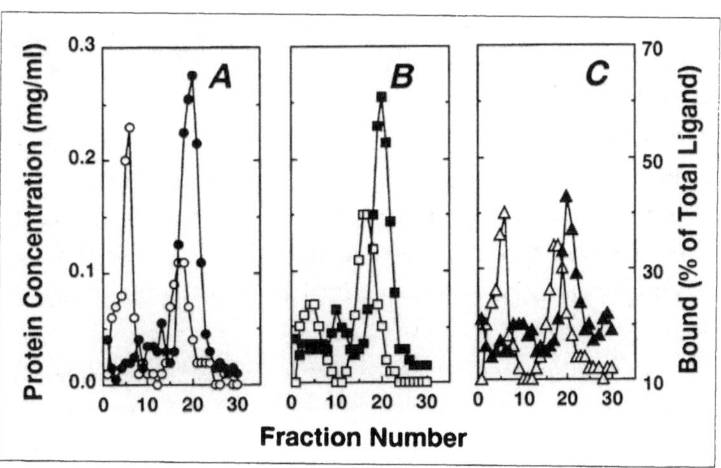

Figure 3.6 Effect of detergents on DEAE-Toyopearl chromatography
The digitonin-solubilized protein (1.5 mg) from brain membranes was subjected to a DEAE-Toyopearl column (packed in a Pharmacia HR5/5 column) chromatography. Elution was performed with a linear gradient increase of NaCl concentration from 0 to 0.5 M in TED buffer, including 0.1% digitonin (A), 0.1% BIGCHAP (B), or 0.3% octylglucoside (C). Total ^{125}I-PACAP27 binding to 10-fold diluted fractions (●,■,▲) and protein concentration (○,□,△) were determined.

Protocol 3.4

DEAE-Toyopearl and hydroxylapatite chromatography

The following chromatography should be performed using a Pharmacia FPLC system.

1. Load the solubilized material from 10 brains (3 g protein/5 l) onto a DEAE-Toyopearl (Tosoh Corporation) column (180 to 200 ml packed in a Pharmacia XK50/100 column) equilibrated with 0.1% digitonin-TED buffer (20 mM Tris, 1 mM EDTA, 0.03% NaN_3, and the protease inhibitors; pH 7.4) at a flow rate of 9 ml/min.
2. Elute the receptor with a linear gradient increase of NaCl concentration from 0 to 500 mM for 240 min at a flow rate of 9 ml/min. Collect 45 ml fractions.
3. Assay fractions for ligand (^{125}I-PACAP27) binding activity at a 200-fold dilution (see Protocol 3.7).
4. Load the peak fractions from DEAE-Toyopearl (700 mg protein/700 ml) onto a hydroxylapatite (HCA100S from KOKEN, or Biogel HTP from Bio-Rad column (150 ml in a Pharmacia XK50/100 column) at a flow rate of 9 ml/min.
5. Elute the PACAP receptor at a flow rate of 9 ml/min with buffer C (20 mM potassium phosphate, 0.1% digitonin, and the protease inhibitors; pH 6.8) and buffer D (600 mM potassium phosphate, 0.1% digitonin, and the protease inhibitors; pH 7.6) as follows: Increase buffer D concentration from 0 to 50% for 75 min, hold it at 50% for 25 min, increase it to 100% immediately, and then hold it at 100% for 100 min. Collect each 45-ml fraction in tubes containing 0.4 ml of 100 mM EDTA (pH 7.0)
6. Assay fractions at a 400-fold dilution and pool the peak fractions.

Affinity Purification of the PACAP Receptor

For retrieval of the PACAP receptor protein, the PACAP receptor was first allowed to form a receptor-biotinylated ligand complex. The amount of the biotinylated ligand required for saturating the PACAP receptor was estimated by a competitive receptor-binding experiment. The estimated concentration corresponds to 10- to 30-

fold of receptor concentration determined from the saturation receptor-binding experiment. The receptor-biotinylated ligand complex was then adsorbed to ImmunoPure® Immobilized Avidin (Pierce). The agarose immobilized 2 mg/ml avidin, which is equal to 30 nmol avidin/ml (120 nmol biotin-binding sites/ml). Finally, the adsorbed PACAP receptor was released from the avidin-agarose column by treatment with acidic buffer (pH 4.0). This elution did not liberate the biotinylated ligand from avidin-agarose, as shown by an experiment using a tracer ^{125}I-biotinylated PACAP27 (Fig. 3.7).

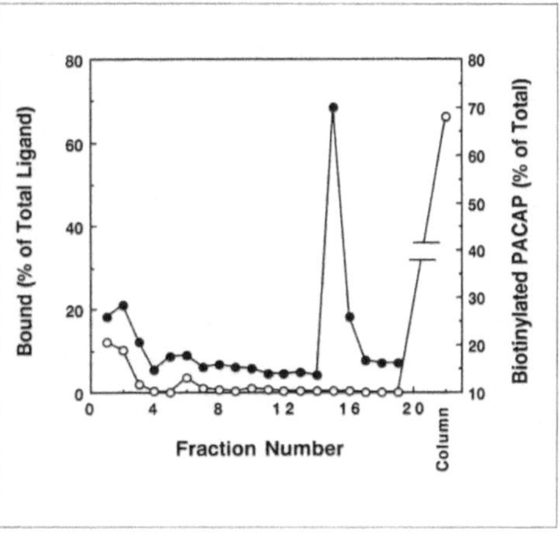

Figure 3.7 Elution of the PACAP receptor but not of biotinylated PACAP from avidin-agarose

The recombinant PACAP receptor expressed in Sf9 insect cells was solubilized with digitonin. The solubilized receptor (150 pmol) was incubated with the biotinylated PACAP (2 nmol) in the presence or absence of tracer ^{125}I-biotinylated PACAP (1 pmol) at 4 °C overnight, and then mixed with avidin-Affigel 10 (0.1 ml). The mixture was rotated gently for 3 days and transferred to a small column. The elution was carried out with a neutral 0.3 M NaCl-, 1 M NaCl buffer and then an acidic 1 M NaCl buffer (pH 4.0). Total ^{125}I-PACAP27 binding to diluted fractions from the experiment without ^{125}I-biotinylated PACAP (●) and total radioactivity in fractions from the experiment with ^{125}I-biotinylated PACAP (○) were determined.

Alternatively, avidin-agarose that preadsorbed the biotinylated ligand could be used for purification. Actually, after the elution with acidic buffer the agarose is reequilibrated with TED buffer and recycled for further adsorption of the receptor. Although the following protocol applied a large excess of the agarose, it could be decreased to one-tenth of the amount below. In our recent procedures for purification of recombinant PACAP receptor, we employed 1 ml of homemade avidin-Affigel 10 immobilizing 2 mg of avidin for 20 nmol of the biotinylated PACAP. There was no significant difference in capacity for the biotinylated PACAP between ImmunoPure® Immobilized Avidin and avidin-Affigel 10; however, the latter allowed better recovery of the PACAP receptor.

Protocol 3.5 **Biotinylated PACAP27-avidin-agarose chromatography**

1. Subject the partially purified PACAP receptor from 10 brains (7 nmol PACAP receptor/400 ml) to repeated cycles of concentration in a Centriprep 10 or Centriplus 10 (Amicon) and dilution with TED buffer for desalting. Finally, concentrate the receptor 4-fold.
2. Add 5 to 10 ml of avidin-agarose (ImmunoPure® Immobilized Avidin from Pierce) to the desalted, concentrated receptor. Agitate the agarose suspension by rotating overnight for adsorption of residual avidin-binding proteins.
3. Transfer the suspension into a column and wash the column with 10 bed-volumes of TED Buffer. Collect all the flow-through (total 150 ml) in polypropylene tubes.
4. Add 220 nmol of biotinylated PACAP (final concentration 1.5 μM) to the flow-through and incubate the mixture at 4 °C overnight to form the receptor-ligand complex.
5. Add avidin-agarose (4000 nmol of immobilized avidin in 130 ml of agarose) to the preformed receptor-ligand complex solution and mix gently with a rotator at 4 °C for 4 days.
6. Sediment the avidin-agarose and transfer it into a column (Econo-Column 5 × 10 cm, Bio-Rad).
7. Wash the column with 10 bed-volumes of 0.3 M NaCl-0.1% digitonin-TED buffer and with 5 bed-volumes of 1.0 M NaCl-0.1% digitonin-TED buffer.
8. Elute the PACAP receptor with a 20 mM magnesium acetate buffer containing 1.0 M NaCl, 0.1% digitonin, 10% glycerol, and the protease inhibitors (pH 4.0). Collect each 10 ml fraction in tubes containing 2.5 ml of 1.0 M Tris (pH 7.5).
9. Assay fractions at a 300-fold dilution and pool the peak fractions.

Final Purification of the PACAP Receptor

The affinity-purified PACAP receptor preparation, which still contained contaminating proteins, was further purified by hydroxylapatite and gel-filtration chromatography.

Protocol 3.6

Hydroxylapatite and Superose 6 chromatography

1. Load the affinity-purified PACAP receptor onto a small hydroxylapatite column (1.6 ml in a Poly-Prep Column, Bio-Rad) at a flow rate of 0.25 ml/min.
2. Elute the receptor with buffer C and buffer D (see Protocol 3.4) at a flow rate of 0.25 ml/min as follows: Increase buffer D concentration from 0 to 50% for 60 min, hold it at 50% for 10 min, increase it to 100% immediately, and then hold it at 100% for 80 min. Collect 1-ml fractions.
3. Assay fractions for ligand-binding activity at a 1000-fold dilution.
4. Concentrate the eluate to 0.1 ml using a Centricon 10 (Amicon) and gel-filter it on a Superose 6 column (1 × 30 cm, Pharmacia) equilibrated with 0.05% digitonin-TED buffer containing 0.2 M NaCl at a flow rate of 0.5 ml/min. Collect 0.5-ml fractions.
5. Assay fractions at a 2000-fold dilution.
6. Store the purified receptor at 4 °C.
   Exchange of digitonin for a more convenient detergent with a higher critical micelle concentration (such as BIGCHAP) could also be performed with hydroxylapatite chromatography. For this purpose, load the purified receptor onto a small column (0.1 ml), wash the column with 10 mM Hepes–0.5 M NaCl buffer (pH 6.8), and then elute the receptor with buffer D containing 0.01–0.1% BIGCHAP instead of digitonin.

Receptor-Binding Assay

The radiolabeled ligand for this assay, ^{125}I-PACAP27, is prepared by radioiodination with lactoperoxidase and subsequent HPLC separation (12). It also is available from Du Pont-NEN. The solution of ^{125}I-PACAP27 should include 0.05% CHAPS and 0.1% BSA; otherwise, it is very difficult to pipette a constant count of the radiolabeled ligand because of its severe adsorption to polypropylene walls. The radiolabeled PACAP27 is usually stored at a 30 to 50 nM concentration in 0.05% CHAPS–0.1% BSA-TED buffer at –30 °C.

Protocol 3.7

Saturation-binding assay

1. Dilute the receptor preparation with 0.05% digitonin-0.1% BSA-TED buffer to a concentration range of 30 to 50 pM – where about 30% of the added ^{125}I-PACAP27 (final concentration = 100 pM) binds to the PACAP receptor.
2. Incubate aliquots of diluted receptor (90 μl) with 10 μl of ^{125}I-PACAP27 (0.2 to 5 nM) in the presence (for determination of nonspecific binding) and absence (for total binding) of 0.3 μM PACAP27 at 25 °C for 75 min.
3. Add 1.5 ml of chilled 0.05% CHAPS-0.1% BSA-TED buffer to the reaction mixtures and filter them through GF/F filters (2.5-cm circle, Whatman) pretreated with 0.03% polyethyleneimine and set on a sampling manifold (Millipore).
4. Wash the filters with 1.5 ml of the same buffer.
5. Count radioactivity trapped on the filters with a gamma-ray counter.
6. Derive specific binding by subtracting nonspecific binding from total binding.
7. Analyze the binding data following Scatchard plot analysis to obtain the dissociation constant and the number of binding sites. The specific activity is defined as the number of binding sites/mg of protein.
   The PACAP receptor in chromatography fractions is assayed as above but with a fixed concentration of ^{125}I-PACAP27 (final concentration = 100 pM). The competitive-binding experiment is done with a fixed concentration of ^{125}I-PACAP27 and increasing concentrations of unlabeled peptide.

3.3 Results and Discussion

The present study reports a method for purifying the PACAP receptor from crude solubilizate. An initial step using DEAE-Toyopearl and subsequent hydroxylapatite chromatography purified the receptor 30-fold with a good yield (Fig. 3.8). The chief virtue of this step, however, was that it almost completely removed a 57K avidin-binding protein that interferes with biotinylated-ligand avidin-agarose affinity chromatography, the next step in the purification.

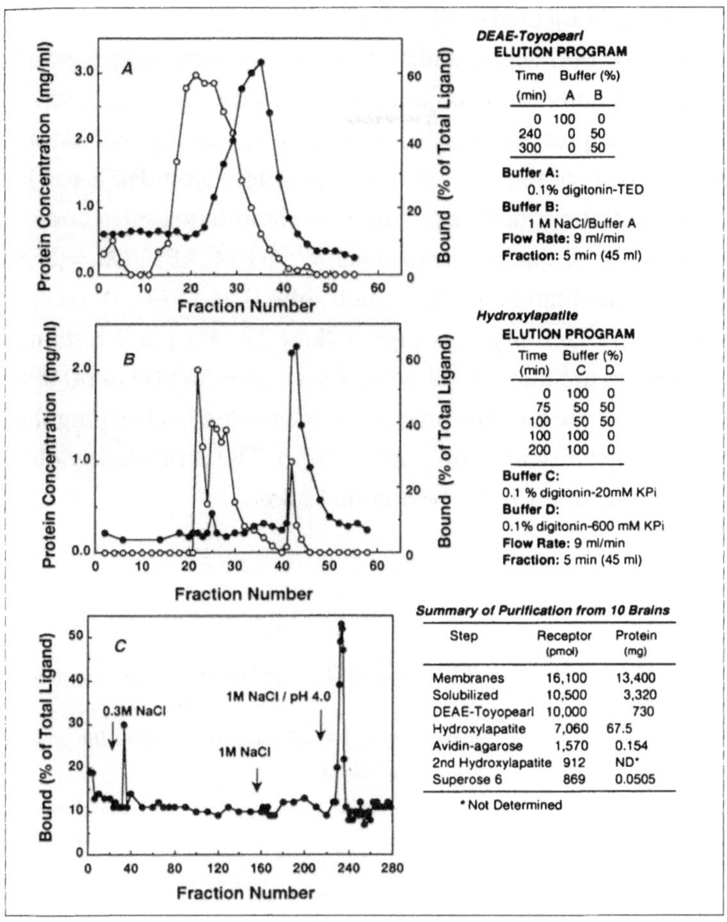

Figure 3.8 Purification of the PACAP receptor

The crude solubilized PACAP receptor from 10 brains was partially purified by DEAE-Toyopearl (A) and hydroxylapatite chromatography (B), and affinity-purified by biotinylated PACAP-avidin-agarose-chromatography (C) as described in Technical Procedures. Total ^{125}I-PACAP27 binding (●) and protein concentration (O) were determined.

As shown in Fig. 3.8, the PACAP receptor bound to the affinity column was not susceptible to elution with 1.0 M NaCl-neutral buffer, but was successfully released by elution with 1.0 M NaCl-acidic buffer (pH 4.0). The yield of this step was not good (20% to 40%), and this step will subsequently require further refinement. Finally, the affinity-purified PACAP receptor was purified further by hydroxylapatite and Superose 6 gel-filtration chromatography.

The purified preparation was subjected to SDS-PAGE. A single broad band with an M_r ranging from 55 000 to 60 000 was visualized by the silver-staining method (Fig. 3.9). Dimer and tetramer bands were sometimes observed even under reducing conditions. The purified receptor was further analyzed by saturation-binding assay. The Scatchard plot analysis of the saturation-binding data provided a single straight line (Fig. 3.9), demonstrating that the purified preparation includes homogeneous ligand-binding sites. The dissociation constant of ^{125}I-PACAP27 (K_d = 25.8 pM) was almost the same as that obtained for the crude solubilized PACAP receptor (K_d = 21.8 pM). The specific activity was higher than 17 000 pmol binding sites/mg protein. This is close to the theoretical value for an M_r = 55 000 to 60 000 protein with a single binding site. Hence, this procedure succeeded in purifying the PACAP receptor to near homogeneity in a fully active form. The procedure yielded about 50 μg of the purified receptor from 10 bovine brains.

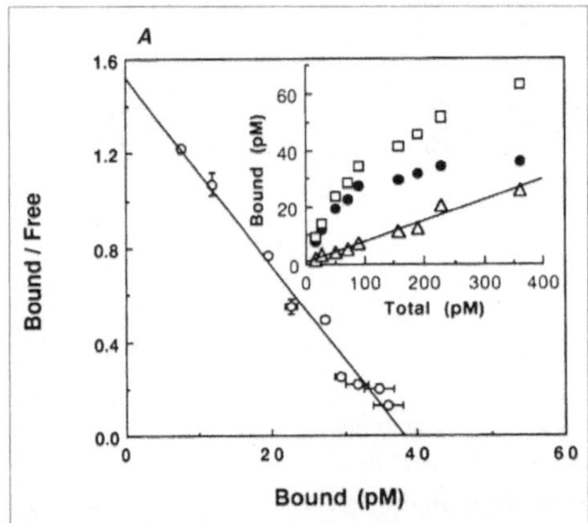

Figure 3.9 Characterization of the purified PACAP receptor

(A) *The purified PACAP receptor was diluted with 0.05% digitonin–0.1% BSA-TED buffer and subjected to a saturation binding assay as described in Technical Procedures. Total binding (□), nonspecific binding (△), and specific binding (●) were plotted versus total ligand (inset). The binding data was analyzed by Scatchard plot analysis (○).*

(B) *The purified PACAP receptor (right lane) and molecular weight markers (left lane) were subjected to SDS-PAG and visualized by the silver-staining method.*

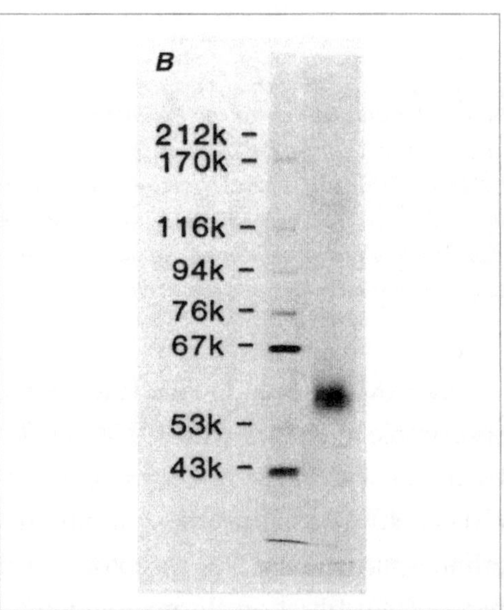

One of the most important points in the present procedure is the successful application of affinity chromatography using biotinylated ligand. The advantage of this chromatography is that the adsorbed PACAP receptor can be eluted under nondenaturing conditions, which is not possible using conventional affinity resins immobilizing PACAP-Cys-NH_2 directly. The reason for this difference is unclear. Moreover, this method decreased nonspecific adsorption to the affinity resins because it applied a minute amount of ligand sufficient to bind most of the receptors.

Another important point revealed in this study is the stabilizing solubilization of the PACAP receptor with digitonin. For most G protein-coupled receptors, including the PACAP receptor, uncoupling from G proteins decreases affinity for agonist ligands. The reduction of ligand-binding affinity often seen in the solubilization step may be caused in part by such uncoupling. The PACAP receptor, however, retained a high affinity for ligand in the solubilized state or in the purified state. The most plausible explanation for this result is that the solubilization with digitonin stabilized the PACAP receptor in this high-affinity form independent of coupling to G proteins.

The present study demonstrates that biotinylated ligand-immobilized avidin affinity chromatography is a highly promising method for purification of the PACAP receptor. The method is not restricted to the PACAP receptor but will be expanded to other ligand-receptor systems or interacting bioactive molecules.

3.4 Troubleshooting

Impurity in PACAP27-Cys-NH_2

In the course of refining this procedure, we encountered several technical difficulties that bear mentioning. Among these were problems due to the insufficient purity of the PACAP ligand. One method for ensuring purity was to synthesize PACAP27-Cys-NH_2 using an automatic peptide synthesizer (Model 430A, Applied Biosystems) and removing of protective groups by the HF method. The deprotected peptide can then be purified by simple gel filtration in a Sephadex G-25 column, but not by reversed-phase HPLC. This procedure did not provide completely pure peptide, but it was effective in preserving the thiol group. Some contaminants in this preparation are not of consequence, because they are removed by the reversed-phase HPLC performed after the biotinylating reaction. However, care should be taken to minimize contaminants. First, the reaction conditions for each coupling cycle should be optimized carefully to complete the coupling reaction. Second, lyophilized PACAP27-Cys-NH_2 powder should be divided into aliquots for each use and kept in a freezer, because thiol-containing peptides are likely to deteriorate. When the biotinylated peptide preparation contains a considerable amount of unbiotinylated contaminants, preadsorption of the biotinylated PACAP to avidin-agarose and subsequent adsorption of the PACAP receptor is also effective.

Low specific activity of the PACAP receptor despite a single band

One possibility is contamination of the 57K avidin-binding protein that co-migrates with the PACAP receptor in SDS-PAGE under reducing conditions. The 57K protein is separated from the PACAP receptor under nonreducing conditions (the PACAP receptor migrates faster than the 57K protein). Another possible cause of low specific activity is the loss of affinity for the ligand during the exchange of digitonin for BIGCHAP. The concentration of BIGCHAP should not rise beyond 100-fold of the receptor concentration. Denaturation of the PACAP receptor by acidification during elution from avidin-agarose is not feasible. Finally, a faulty protein determination may result in an aberrantly low specific activity. Protein concentration should be determined by the method described by Schaffner and Weissmann (13). The Bradford method overestimates the protein concentration of the detergent-containing samples.

Acknowledgments

We are grateful to Drs Masahiko Fujino, Hisayoshi Okazaki, Kyozo Tsukamoto, Takehiko Naka, and Masao Tsuda for their discussions and encouragement throughout this work.

References

1 Finn, F.M., Titus, G., Horstman, D. and Hofmann, K. (1984) *Proc. Natl. Acad. Sci. USA* **81**, 7328–7332.

2 Redeuilh, G., Secco, C. and Baulieu, E.E. (1985) *J. Biol. Chem.* **260**, 3996–4002.

3 Hazum, E., Schvartz, I., Waksman, Y. and Keinan, D. (1986) *J. Biol. Chem.* **261**, 13043–13048.

4 Finn, F.M. and Hofmann, K. (1990) *Methods Enzymol.* **184**, 244–274.

5 Redeuilh, G., Secco, C. and Baulieu, E.E. (1990) *Methods Enzymol.* **184**, 292–300.

6 Kozuka, M., Ito, T., Hirose, S., Lodhi, K.M. and Hagiwara, H. (1991) *J. Biol. Chem.* **266**, 16892–16896.

7 Eppler, C.M., Zysk, J.R., Corbett, M. and Shieh, H.-M. (1992) *J. Biol. Chem.* **267**, 15603–15612.

8 Brown, P.J. and Schonbrunn, A. (1993) *J. Biol. Chem.* **268**, 6668–6676.

9 Ohtaki, T., Masuda, Y., Ishibashi, Y., Kitada, C., Arimura, A. and Fujino, M. (1993) *J. Biol. Chem.* **268**, 26650–26657.

10 Schäfer, H. and Schmidt, W.E. (1993) *Eur. J. Biochem.* **217**, 823–830.

11 Murphy, J.B. and Kies, M.W. (1960) *Biochim. Biophys. Acta* **45**, 382–384.

12 Ohtaki, T., Watanabe, T., Ishibashi, Y., Kitada, C., Tsuda, M., Gottschall, P. E., Arimura, A. and Fujino, M. (1990) *Biochem. Biophys. Res. Commun.* **171**, 838–844.

13 Schaffner, W. and Weissmann, C. (1973) *Anal. Biochem.* **56**, 502–514.

Photoreactive Biotinylated Peptide Ligands for Affinity Labeling

M. Fabry and D. Brandenburg

Summary

The combination of photoaffinity labeling with reversible biotin-avidin complex formation, a method termed PAMAC (photoaffinity-mediated avidin complexing), is a powerful approach in the analysis of ligand-receptor systems that combines the advantages of both techniques. Ligands are designed either for permanent or temporary biotinylation and are – if necessary after additional radiolabeling – photocross-linked to receptors. Biotinylated irreversible conjugates or subsequent enzymatic fragments can be isolated by avidin-streptavidin affinity chromatography. Temporary biotinylation allows isolation of covalent ligand-receptor complexes after cleavage of the biotin group. With permanent biotinylation, all preparative and analytical advantages of the biotin label are preserved, for instance, high-sensivity detection by enhanced chemiluminescence, or dot-staining procedures.

Using insulin as an example, practical procedures for synthesis of a trifunctional labeling reagent, 2 photo/biotin-labeled hormone derivatives, and their applications in receptor binding-site analysis are described in detail. Brief reference is made to related work with other peptide hormones.

4.1 Introduction

Photoaffinity labeling is one of the major techniques for obtaining information on interacting biological systems (1). Photoreactive peptide and protein hormones have played an important role in the identification of receptors and in investigations of their structure and function (2, 3).

Similarly, biotin-avidin/streptavidin systems have proved their power and almost unlimited utility in many areas of basic and applied biochemistry (4). The combination of both techniques appeared as a particularly promising approach for refined analyses of ligand-receptor systems. The concept was introduced in 1985 by Finn and Hofmann (5) with ACTH. We have applied it to insulin, and termed it PAMAC – photoaffinity-mediated avidin complexing (6).

Photoaffinity labeling generally involves reversible complexing of a photoreactive ligand with the receptor, generation of a highly reactive intermediate by irradiation, and formation of a covalent, irreversible bond between hormone and receptor. The presence of a biotin moiety in the ligand subsequently allows both the selective "fishing" of the covalent complex and its specific detection via avidin-linked detection systems. For many applications, the position of either label in the ligand molecule may be of minor importance, and will only be governed by the requirement of minimal interference with receptor binding. For a variety of detailed studies, however, the site of both labels is deliberately determined. There are principally 2 strategies, which combine photo and biotin labeling in different ways: temporary biotinylation of photoreactive molecules, and permanent biotinylation.

Accordingly, there are 2 types of labeled ligands: In the first approach, ligands are tagged at 2 different, independent sites with photo and biotin labels. This can be a matter of design (see cleavable biotin labels, below) or may depend on structure and/or feasibility. In the second approach, photo and biotin labels are placed in the immediate vicinity and should be chemically or enzymatically nonseparable. This can be achieved in 2 ways:

(i) a premade substituent containing both groups is linked to the complete, and, if necessary, partially protected ligand, or is used as a building block in the course of ligand synthesis, or (ii) both features are incorporated into the ligand in a stepwise manner. In many cases a further requirement is the presence of a radioactive label in the ligand, which poses additional preparative constraints. However, the recent development of chemiluminescence as a highly sensitive analytical tool makes this last requirement increasingly superfluous. The use of biotin-avidin complexing in chemiluminescence assays is a further extremely valuable extension of its applicability (7).

Examples from our work with insulin will be presented to illustrate the synthesis of a trifunctional reagent, labeling of ligands using permanent biotinylation, and applications in the analysis of the receptor. Some other applications will be briefly discussed. For a previous review, see Brandenburg et al. 1990 (3).

4.2 Technical Procedures

Synthesis of a trifunctional photoactivatable biotinylating reagent

Procedure 4.1

N^{α}-(4-azido-2-nitrophenyl)-biocytin-4-nitrophenyl ester (Nap-Bct-ONp, Fig. 4.1, I) (8)

Biocytin [0.8 g (2.15 mmol), prepared according to ref. 9] was reacted with 0.4 g of 2-nitro-4-azido-fluorobenzene (F-Nap) in 30 ml of ethanol/0.1 M Na_2CO_3 at 60 °C for 72 hours. The solution was concentrated in vacuo and extracted with ethyl acetate. Acidification of the aqueous phase gave an orange precipitate, which was recrystallized twice from methanol/ethyl ether. Yield: 0.5 g (43.5% of theory), mp 94–96 °C (decomp.), 98.8% pure according to RP-HPLC (triethylammonium phosphate/acetonitrile, pH 4.0). ^{1}H-NMR spectra were in agreement with expected values. $\tau_{1/2}$ of Nap-Bct-OH was 84 sec under the irradiation conditions generally used.
The nitrophenyl ester was obtained by reacting 0.5 g (0.94 mmol) of Nap-Bct-OH with 160 mg (1.15 mmol) of 4-nitrophenol and 200 mg (0.97 mmol) of dicyclohexylcarbodiimide (DCC) in 50 ml of dimethylformamide (DMF) at 0–2 °C. Yield: 414 mg (67.2% of theory) of orange powder (from methanol/ethyl ether), mp 106–109 °C (decomp.), homogeneous in TLC.
The synthesis of N^{α}-(4-azidotetrafluorobenzoyl)-biocytin-N-hydroxysuccinimide ester (Atf-Bct-OSu) (Fig. 4.1, II) was prepared in a similar manner. This compound is commercially available from Boehringer Mannheim.

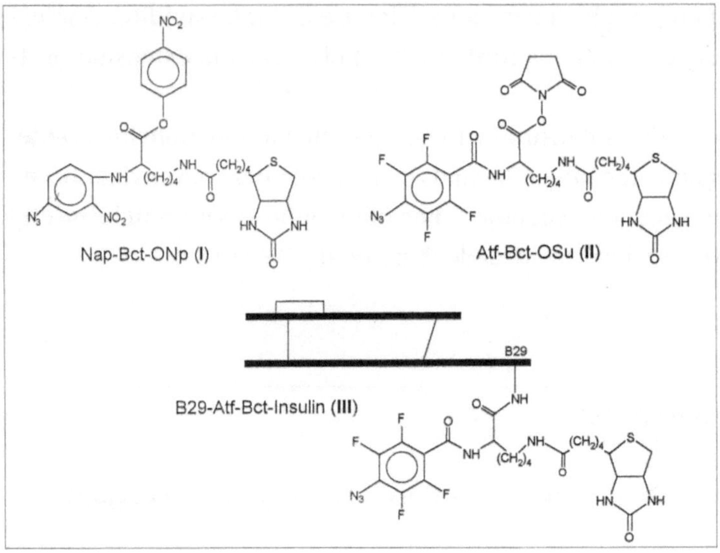

Figure 4.1 Structure of the photobiotin reagents I and II, as well as the insulin derivative III

Synthesis of photoreactive biotinylated peptide hormones

General conditions

The peptide/protein is dissolved at a concentration of 1–20 mg/ml (depending on solubility) in DMF or an aqueous amine-free buffer [e.g., phosphate-buffered saline (PBS)]. Atf-Bct-OSu, dissolved in distilled DMF or dimethyl sulfoxide (DMSO) (1–10 mg/ml), is added in 2 to 10-fold molar excess. The mixture is stirred in the dark for 2–4 hours. The pH should be controlled (pH 7.5–8.5). Excess reagent/hydrolyzed free acid can be separated by gel filtration, membrane filtration, or dialysis.

Procedure 4.2

Synthesis of $N^{\varepsilon B29}$-(4-azidotetrafluorobenzoyl-biocytinyl)-insulin (Atf-Bct-insulin, Fig. 4.1, III)

Ten milligrams (1.7 µmol) of porcine $N^{\alpha A1}$,$N^{\alpha B1}$-bis-Msc-insulin (10) was reacted with 3.4 mg (5 µmol) of Atf-Bct-OSu in 0.33 ml of DMF/*N*-methylmorpholine (1/0.01, v/v) at room temperature for 3 hours under slow shaking. The solution was acidified with acetic acid and subjected to gel filtration on Sephadex G-25 in 10% acetic acid. The protein fraction was lyophilized. The crude material was dissolved in 2.5 ml of 10% aqueous piperidine at 0 °C and kept for

110 min in the dark. The solution was acidified with acetic acid and subjected to gel filtration on Sephadex G-25 in 10% acetic acid. The protein fraction was lyophilized. Yield: 6.4 mg, containing 43% acylation product according to RP-HPLC. This material is well suited for photoaffinity labeling of receptors. If desired, it can be purified by RP-HPLC or avidin affinity chromatography.

Purification of Atf-Bct-insulin by avidin affinity chromatography

The lyophilized reaction product was dissolved in PBS (0.1 M sodium phosphate, 0.15 M sodium chloride, pH 7.2) and applied to a monomeric avidin-agarose column (Pierce). The column was washed with PBS and then with PBS containing biotin (2 mM) to elute the biotinylated photoinsulin. Fractions of 1 ml were collected and monitored for protein at 280 nm. Fractions eluted with biotin were pooled and lyophilized after gel filtration on a PD-10 column (Pharmacia).

Site-specific incorporation of biotin and photo labels in separate steps

Procedure 4.3

4-Azidosalicylic acid-*N*-hydroxysuccinimide ester (Asa-OSu) (11)

4-Azidosalicylic acid was prepared from 4-aminosalicylic acid by diazotation and reaction with NaN_3 essentially according to ref. 12. Yield, after recrystallization from ethanol and ethanol/water: 56%, mp 193 °C. Conversion into the active ester gave 64% of yellow-brownish crystals (from dioxane, then dioxane/petroleum ether), mp 158 °C. Both compounds were homogeneous according to thin-layer chromatography in 3 systems and RP-HPLC (R_t 35.4 min) and showed an IR spectrum with a strong azide band at 2123 cm^{-1}. UV spectrum of Asa-OH: ε_{264} = 13 800 M^{-1} cm^{-1} (50 mM NH_4HCO_3, pH 8.2).

Procedure 4.4

A1,B29-Msc_2-des-(Phe^{B1},Val^{B2})-insulin (II)

Compound II was prepared by 2 cycles of Edman degradation of A1,B29-Msc_2-insulin (I) as described (13). Briefly, 720 mg (120 mmol) of I was reacted with 0.6 ml (4.29 mmol) of phenylisothiocyanate in 60 ml of deaerated pyridine/water (9:1) for 3 hours at room temperature in the dark. The isolated derivative was washed with methanol/ether, dried, and treated with 18 ml of trifluoroacetic acid (TFA) for 1.5 hours. The dried product was resolved in 60 ml of 90% pyridine, and B2-valine was cleaved by treatment under the same conditions. Crude II was purified by ion-exchange chromatography on SP-Trisacryl M (Merck) at pH 3 as above. The yield after gel filtration was 342 mg (50%, based on I).

Procedure 4.5

A1,B29-Msc$_2$-des-Phe^{B1}-[Lys(Msc)^{B2}]insulin (III)

73.1 mg (170.5 μmol) of Boc-Lys(Msc)-OH (see Fig. 4.5, K. Goyal, unpublished) was preactivated by reaction with 26.1 mg (170.5 μmol) of 1-hydroxybenzotriazole (HOBt) and 31.7 mg (153.5 μmol) of dicyclohexylcarbodiimide (DCC) in 0.5 ml of DMF for 40 min at room temperature. This solution was then added to 100 mg of II in a mixture of 1 ml of DMF and 20 μl of *N*-methylmorpholine. After stirring for 3 hours, the solution was filtered, and *N*-protected III precipitated with ether/methanol, washed with ether, and dried. *t*-Butyloxycarbonyl (Boc) groups were cleaved with 3 ml of trifluoroacetic acid; III was precipitated with ether and dried. Yield after gel filtration and lyophilization: 97.1 mg (92.7%, based on II).

Procedure 4.6

A1,B29-Msc$_2$-[Bct^{B1},Lys(Msc)^{B2}]insulin (IV)

76.8 mg (162.8 μmol) of Boc-Bct-OH (8) was preactivated with 24.6 mg (162.8 μmol) of HOBt and 30.2 mg (146.4 μmol) of DCC as before. After 1 hour the solution was added to 100 mg of III (16.3 μmol) in 1 ml of DMF and 20 μl of *N*-methylmorpholine, and the mixture was stirred for 1 hour. After isolation and deprotection under conditions as above, the yield of IV was 95.4 mg (89.7%, based on III). Amino acid analysis: Val 2.73 (3), Phe 1.87 (2), Lys 2.83 (3), based on Glu = 7.0. All other values were in the range expected.

Procedure 4.7

B1-Asa-[Bct^{B1},Lys^{B2}]insulin (V) (Asa-Bct-insulin)

30 mg (4.59 μmol) of IV in 0.9 ml of DMF was acylated with 12.7 mg (45.9 μmol) of Asa-OSu in the presence of 7.03 mg (45.9 μmol) of HOBt for 3 hours at room temperature. The product was isolated by

precipitation, dried, and dissolved in 1.2 ml of DMF/water (1:1). After cooling to 0 °C, 300 μl of cold 2 *N* NaOH/MeOH (2:1) was added. The mixture was stirred in an ice bath for 75 sec and then acidified with 375 μl of glacial acetic acid. Chromatography on Sephadex G-50f in 10% acetic acid gave 22.8 mg (75.2%, based on IV) of crude V. Amino acid analysis: Val 2.72 (3), Phe 1.96 (2), Lys 2.99 (3), based on Gly = 4.0. All other values were in the range expected.

Procedure 4.8

Radioiodination of B1-Asa-[Bct^{B1},Lys^{B2}]insulin (Asa-Bct-insulin)

Asa-Bct-insulin was purified by preparative RP-HPLC (Nucleosil RP-C18, 5-μm column, 250 × 10 mm, 31.5% acetonitrile/0.07% TFA, flow rate 2 ml/min, yield 18%). For iodination 200 μg of homogeneous Asa-Bct-insulin (according to analytical RP-HPLC) was dissolved in 200 μl of phosphate buffer (pH 7.8), 6 M urea (4.595 g of Na_2HPO_4 × 2 H_2O, 0.575 g of KH_2PO_4 dissolved in 87.5 ml of 7 M urea, fill up with distilled water ad 100 ml). The iodination was carried out in the presence of urea in order to increase the extent of the *B*-chain iodination. To compensate for the missing urea in the alkaline $Na^{125}I$ solution, 100 μl of the insulin solution was added to another 18 mg of urea. Twenty microliters of this solution was added to 1 mCi of $Na^{125}I$. The $Na^{125}I$ was dissolved in NaOH, pH 7–11, in a concentration of 3.7 × 10^9 Bq/ml. The iodination was carried out at room temperature with stirring by addition of 2.5 μl of H_2O_2 (0.3 mM, 15.5 μl of 30% H_2O_2 in 5 ml of phosphate/urea buffer, from this solution 2.5 μl in 2.5 ml of phosphate/urea buffer) and 2.5 μl of lactoperoxidase (0.2 mg/ml of phosphate/urea buffer). After 1 min another 2.5 μl of each solution was added. The incorporation of the iodine in insulin was checked after 5 min by precipitation with 10% trichloroacetic acid. If the rate of incorporation is at least 75%, the reaction mixture can be purified either by gel filtration on Sephadex G-50f (Pharmacia) in 50 mM NH_4HCO_3/0.1% BSA, SepPak C-18 cartridge in water/acetonitrile (Millipore) or RP-HPLC. The B1-iodo derivative of Asa-Bct-insulin was isolated by RP-HPLC on a column (250 mm × 4 mm) of LiChrosorb RP-18 (Merck) in 25% acetonitrile 0.25 M triethylammonium phosphate, pH 4.0, flow rate 1 ml/min. Fractions of 1 ml were collected in a solution of 100 μl of 50 mM NH_4HCO_3/1% BSA, counted for radioactivity, and stored at –20 °C.

In the example, a partially protected insulin derivative was used to ensure single-site modification at B29 or B1. The use of unprotected insulin leads to a mixture of labeled species, which, however, can be used for qualitative labeling.

Photoaffinity labeling

General conditions

Intact cells, membranes, or solubilized receptor (partially or highly purified) are incubated with the photoreactive ligand in the dark. Cells and membranes can easily be washed before irradiation (cool to 4 °C, centrifuge for 10 min at 13 000 rpm, resuspend in ice-cold buffer). The mixture is then photolysed to form the covalent ligand-receptor complex. Instead of the UV flash apparatus ("Lizzy," Raytest, D-75334 Straubenhardt), continuous UV lamps can be used (irradiation at 258 nm for 5–20 min). Nonbound ligand is separated from the covalent ligand-receptor complex by gel filtration, membrane filtration, polyethylene glycol precipitation, or SDS-polyacrylamide gel electrophoresis (PAGE). The complex can be analyzed and detected by radioactivity and/or biotin/streptavidin techniques.

Procedure 4.9

Analytical photoaffinity labeling

Examples for a radioactive and a nonradioactive ligand are given. 1 µg of insulin receptor ectodomain is incubated with 0.6–6 nM ^{125}I-Asa-Bct-insulin (2×10^5-2×10^6 cpm) in 100 µl of KRH buffer [(50 mM Hepes, pH 8.3, 130 mM NaCl, 5.1 mM KCl, 1.3 mM $MgSO_4$, 0.1% bovine serum albumin (BSA)] or 40–400 nM Atf-Bct-insulin in 20 µl of PBS for 90 min at room temperature or 16 hours at 4 °C in the dark. The mixture is irradiated with 3 flashes (each 1000 Wsec). Noncovalently bound Asa-Bct-insulin is removed either by precipitation with gamma globulin and polyethylene glycol or by Centricon-30 (Amicon) centrifugation.

Procedure 4.10

Analytical detection of biotin-labeled receptor by Western blotting (Fig. 4.2)

Aliquots of the labeled receptor are separated in a 7.5% SDS-PA gel (for the reduced complex) and in a 4.5% SDS-PA gel (for the whole receptor). Proteins are transferred to a PVDF-membrane by electroblotting. Biotin-containing bands are visualized using a chemiluminescence detection reagent (Boehringer Mannheim). The overall process can be separated into 3 basis steps (for further details, see Hoeltke, this volume):

1. Blocking of all nonspecific binding sites on the membran (30 min)
2. Incubation with a streptavidin-horseradish peroxidase conjugate or anti-biotin-peroxidase, Fab-fragments, and washing of the membrane (30 min, 4 × 10 min)
3. Detection of the label by enzyme reaction leading to immediate emission of light; exposure of X-ray film (exposure sec, min or hours)

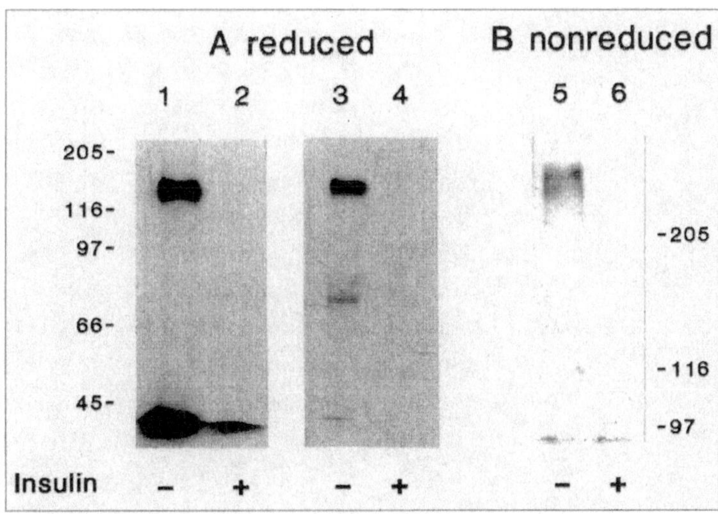

Figure 4.2 Detection of photoaffinity-labeled insulin receptor ectodomain by enhanced chemiluminescence Western blotting
Two picomoles of ectodomain (lanes 1, 2, 5, and 6) and WGA-purified insulin receptor from NIH3T3 fibroblasts (lanes 3 and 4) were incubated with 20 pmol (lanes 1 and 2), or 8 pmol (lanes 3–6) of Atf-Bct-insulin for 60 min in the ab-

sence (lanes 1, 3, and 5) and the presence (lanes 2, 4, and 6) of 1000-fold excess unlabeled insulin (final volume 20 μl). The samples were photoaffinity labeled by UV irradiation, resolved in a reducing [100 mM DTE, (A) 7.5%] and a nonreducing [(B) 4.5 %] SDS-PAGE, and transferred to a PVDF membrane by electroblotting. The blot was probed with horseradish peroxidase-conjugated streptavidin, and enhanced chemiluminescence was performed. The initial amount of receptor subjected to the SDS-PAGE was (1, 2) 100 ng in (A) and 40 ng in (B). Exposure time of the X-ray film was 5 min. In the presence of reductant (A) a biotin-positive band is observed at 123 kDa, as expected for the covalently labeled α subunit of the receptor (11). In (B) the nonreduced receptor appears as 2 bands above 350 kDa. The protein weight markers used were 205 kDa myosin, 116 kDa β-galactosidase, 97 kDa phosphorylase, 68 kDa BSA, 45 kDa ovalbumin. The ectodomain was a gift from Drs Erik Schaefer and Leland Ellis.

Procedure 4.11

Preparative photoaffinity labeling, tryptic digestion, and isolation of labeled receptor fragments by streptavidin affinity chromatography (14)

Insulin receptor ectodomain (400 μg) is incubated in 50 mM Tris-HCl buffer, pH 8.0, with Asa-Bct- and ^{125}I-Asa-Bct-insulin (receptor:insulin = 1:1.5) for 16 hours at 4 °C. Irradiation (6 flashes) is carried out in a petri dish under gassing with helium. Noncovalently bound insulin derivative is removed by Centricon-30 centrifugation and washing 3 times with 2 ml of 50 mM Tris-HCl buffer, pH 8.0. Digestion is performed with 30 μg/ml TPCK-treated trypsin (Merck) for 24 hours at room temperature in buffer containing 50 mM Tris and 10 mM $CaCl_2$, pH 8.0, and is stopped by addition of 2 equivalents of trypsin inhibitor (alternatively pH lowering or heating). Streptavidin-agarose (0.8 ml, 1.2 mg of streptavidin per ml of packed gel, Boehringer Mannheim) is washed with 45 volumes each of 0.1 M $NaHCO_3$, 1 M NaCl, pH 8.0, and then 0.1 M sodium acetate, 1 M NaCl, pH 4.0, and equilibrated with 45 volumes of buffer containing 50 mM Hepes and 1 mM EDTA, pH 8.2. The tryptic digest is diluted with 1.5 ml of Hepes buffer and incubated with the streptavidin-agarose for 16 hours at 4 °C under rotation. The resin is filled into a column and washed with 200 volumes of 50 mM Hepes, 1 M NaCl, and, as detergent, 0.1% polyoxyethylene-10-

tridecyl ether (Sigma), pH 7.6. The biotinylated peptides are eluted by repeated (5 times) heating with 0.3 ml of 50 mM Hepes, 1 mM EDTA, and 1% SDS, pH 8.3, at 95 °C for 15 min until radioactivity in the eluates is close to basal values. The eluates are pooled, concentrated by membrane ultrafiltration (Ultrafree-MC NMWL 5000, Millipore), and applied to the HPLC column (Fig. 4.3). The tubes are screened for biotin content, and the material from the major peaks is isolated by SpeedVac lyophilization.

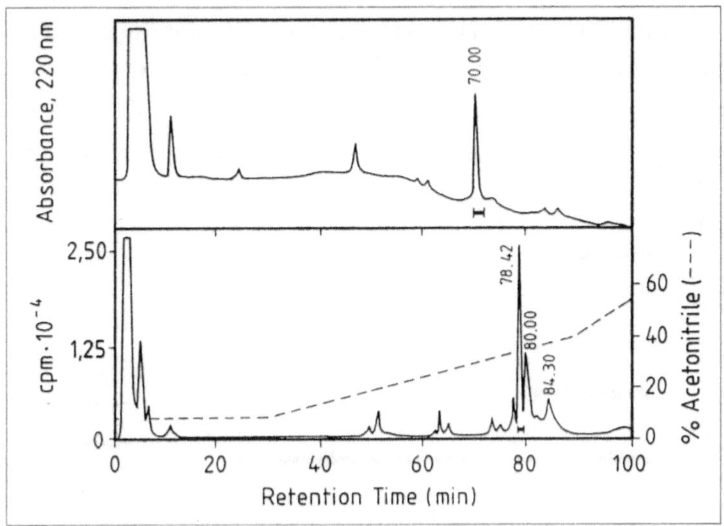

***Figure 4.3* RP-HPLC purification of the tryptic peptides of photobiotin-labeled insulin receptor ectodomain isolated by streptavidin affinity chromatography**

Ectodomain (240 nM) was labeled with 0.17 nM B1-^{125}I-Asa-Bct-insulin and 360 nM Asa-Bct-Insulin (molar ratio of receptor to insulin 1.0:1.5). Noncovalently bound insulin derivative was removed by Centricon-30 centrifugation (Amicon). After tryptic digestion (30 µg/ml of TPCK-trypsin, 24 hours, 37 °C) and addition of trypsin inhibitor the biotinylated receptor fragments were isolated by streptavidin affinity chromatography following RP-HPLC purification. The tubes were screened for biotin in a dot blot (avidin biotinylated-horseradish peroxidase and 4-chloro-1-naphthol, acc. to ref. 4, chapter 50) and the material from the major peaks (fractions 72–75 based on UV and fractions 82–85 based on radioactivity) was isolated by SpeedVac lyophilization. From Fabry et al. (14) with permission.

4.3 Results and Discussion

Synthesis of photoactivatable insulins with permanent biotin labels

Lysine appeared to be a suitable scaffold to carry the photo and biotin group, ensuring sufficient stability, flexibility, and geometry for avidin interaction. In Nap-Bct the distance between the nitrene nitrogen and the biotin ring is 1.9×10^{-9} cm. The perfluoro derivative should be favourable, because irradiation gives nitrenes with greater stability than with normal azides (15). The active esters are thus trifunctional reagents for one-step protein labeling with both groups.

For the fingerprint analysis of the insulin-binding site of the receptor, further requirements were (Fig. 4.4) an adjacent site for radioactive labeling (Tyr^{B26}), and a site for enzymatic cleavage (Arg^{B22} for trypsin, Glu^{B21} for V8-protease). Hence, the side chain of Lys^{B29} was an appropriate position for labeling.

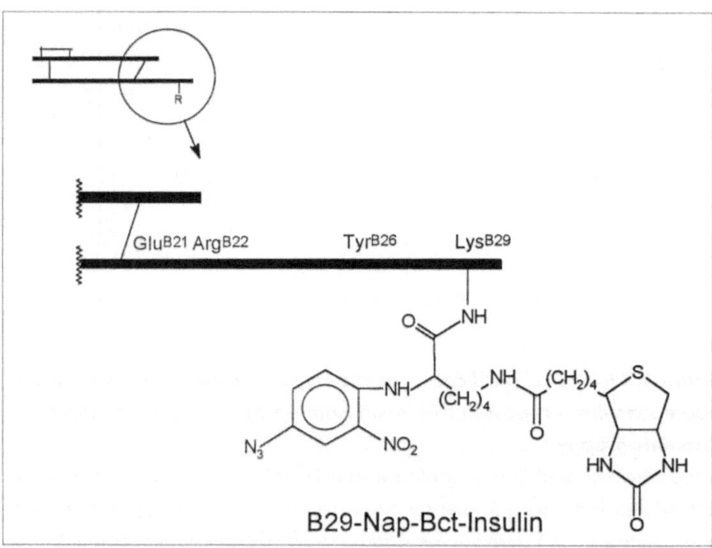

Figure 4.4 Structure of B29-Nap-Bct-insulin

Labeling at the N-terminus and generation of a new tryptic cleavage site proceeded as follows (Fig. 4.5):

1. Preparation of A1,B29-diprotected Msc-insulin
2. Two cycles of Edman degradation to remove Phe and Val
3. Acylation at B3 with Boc-Lys(Msc)-OH, deblocking at B2

4. Acylation with Boc-Bct-OH, deblocking at B1
5. Acylation with azidosalicylic acid
6. Removal of all 3 Msc groups

Because simultaneous photolabeling and iodination with Asa-OSu gave only low yields, it was necessary to iodinate the Asa-insulin and to separate the iodo isomers by HPLC.

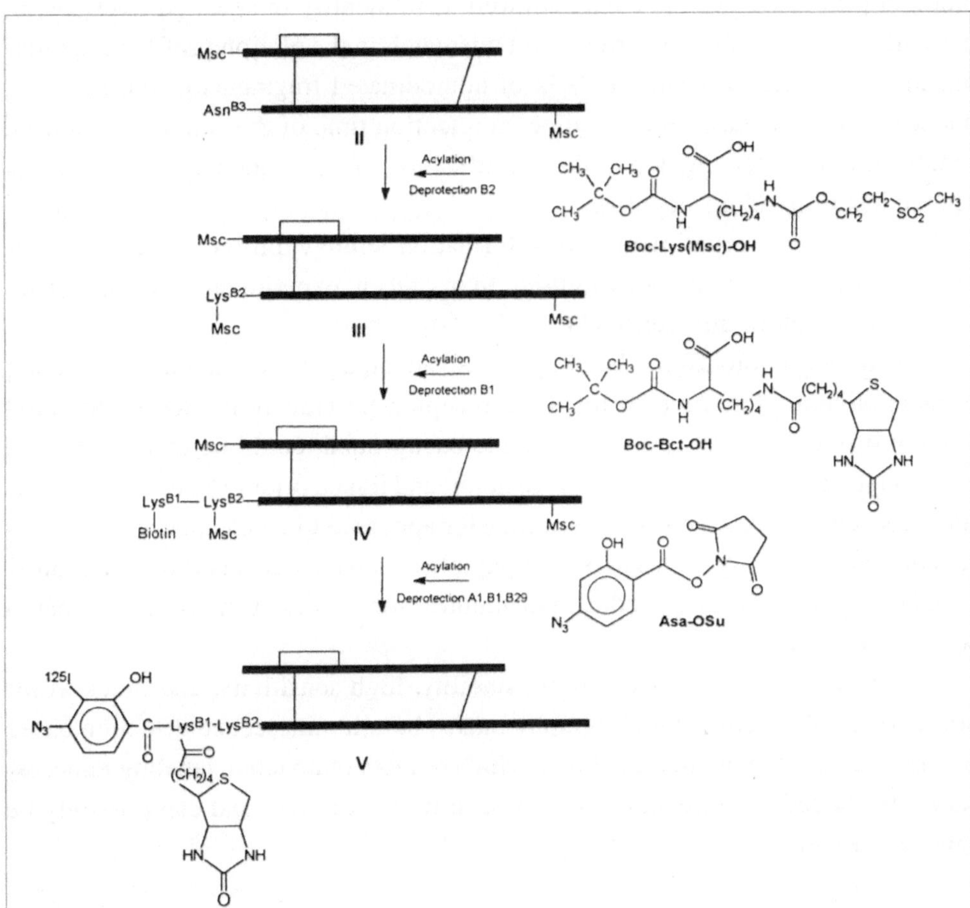

***Figure 4.5** Synthesis of B1-Asa-[Bct^{B1},Lys^{B2}]insulin, shown in its final radioiodinated form*

Applications: Insulin

We have used B1-^{125}I-Asa-Bct-insulin, as in previous work with the B29-Nap-Bct-insulin, in the analysis of the hormone-binding site. After covalent coupling of the insulin derivative to the receptor, the biotin residue served as a handle for the isolation of conjugated fragments after enzymatic digestion by affinity chromatography. As observed earlier, desorption necessitated rather harsh conditions but was nevertheless satisfactory. In the subsequent HPLC separation of the fragments, a biotin dot-blot assay of the eluate allowed us to identify the desired fractions. In particular, we were able to correct for a mismatch in the elution profile of the iodinated tracer fragments and the bulk of noniodinated fragments of the same sequence caused by the effect of iodine on retention time of the smaller fragments. Finally, microsequencing led to the identification of the sequence α390–470 as a second insulin-binding site (14).

In these earlier experiments, color formation with avidin-biotinylated peroxidase and 4-chloro-1-naphthol was used. In our recent experiments, we applied detection via chemiluminescence (1).

In analytical photoaffinity labeling several advantages over radioactive labeling became obvious. The ratio of ligand to receptor, previously limited by the total amount of radioactivity and unfavourable background, could be increased up to 20:1, yet it was not necessary to remove nonbound ligand prior to SDS-PAGE. Blotting proceeded with high resolution; transfer appeared to be cleaner and more efficient. The sensitivity seems to be comparable to carrier-free labeling with iodine-125, but exposure times are reduced to minutes, sometimes even less, as compared with several days.

Conclusion: Convenient handling, stability, high sensitivity, and quick availability of results from autoradiography make chemiluminescence detection based on biotin/avidin the method of choice which renders radioactive labeling unnecessary. The detection limits have not yet been fully explored, and can possibily be lowered further.

Examples for other applications

The synthesis of new ligands for single-site, permanent biotinylation in combination with photoaffinity labeling has been described:

N^{α}-Biotin-4-azido-phenylalanin-N-hydroxysuccinimide ester was used for simultaneous labeling of an LH-RH analogue at the side chain of D-Lys^4 (16). Similarly, a $glycine^9 \rightarrow$ lysine replacement in arginine-vasopressin (AVP) provided a site for attachment of a group containing biotin and an iodinatable photo group (17). Besides radiolabeling and some preliminary photoaffinity labeling, these derivatives have not yet been used in receptor studies, particularly not using biotin complexing.

Cleavable biotin anchors are useful for temporary biotinylation in combination with photoaffinity labeling. Several derivatives of angiotensin have been synthesized that contain biotin or iminobiotin, linked directly or via spacer molecules to the N-terminus of angiotensin II, and 4-azido-phenylalanin in the C-terminal position 8 (18). Ahmed et al. (19) attached biotin or a cleavable biotin derivative to the N-terminus, and a photoreactive diazirine group to Lys^{11} close to the C-terminus of α-MSH. Using the reagent Biot-NH-$(CH_2)_2$-SS-$(CH_2)_2$-CO-OSu(SO_3Na) = Biot-SS-OSu, which is now commercially available (Pierce), 2 derivatives were obtained: angiotensin II, Biot-NH-$(CH_2)_2$-SS-$(CH_2)_2$-CO-A-R-V-Y-I-H-P-F(N_3) (18), and α-MSH, Biot-SS-S-Y-S-N-E-H-f-R-W-G-K(photolabel)-P-V-NH_2 (19). Both iodinated derivatives were used to affinity-label the corresponding receptors and to isolate the biotinylated covalent complexes, either by affinity chromatography (20) or on magnetic beads (19). Reductive cleavage with DTT/SDS cleaved the biotin moiety and liberated the hormone-receptor conjugates in very high yield. The AII receptor could be further analyzed for the AII-binding site (21).

4.4 Troubleshooting

The reaction with a protein should be carried out with a fresh solution of the biotinylated photoreactive ester and freshly distilled solvents or amine-free buffers. The pH of the reaction must be controlled and should be kept between 7.5 and 8.5. More active ester can be added if the reaction proceeds only slowly. The reaction volume should be kept minimal, because in concentrated solutions acylation is favoured over hydrolysis.

Deblocking of Msc groups is milder with piperidine than with NaOH, but temperature should also be controlled well.

Since the introduction of substituents may affect the interaction of the ligand with target molecules, binding studies under reversible conditions have to be carried out in order to assess the applicability of labeled ligand. If the modified ligand does not bind to the receptor, pH may need to be corrected. It should be adjusted to the optimum of the interaction between ligand and receptor. Buffer and incubation time should also be optima (e.g., for insulin 15–60 min at 30–37 °C, or 5–16 hours at 4 °C). BSA should always be included in a concentration of 1–10 mg/ml to reduce adsorption of the proteins to the surface of plastic vials and especially to glass. Some ligands, however, bind to BSA or to other proteins in an unspecific manner, particularly if the amount of photoreactive ligand is too high.

The dissociation of the tetrameric avidin or streptavidin-biotin complex requires harsh conditions (8 M guanidine, pH 1.5, heating in SDS-electrophoresis sample buffer). However, monomeric Avidin or iminobiotin require milder dissociation conditions.

A problem in various experimental systems is the presense of endogenous biotin. This can be overcome by preclearing before labeling, for instance, through partial purification on WGA or on immobilized streptavidin.

References

1 Brunner, J. (1993) *Annu. Rev. Biochem.* **62**, 483–514.

2 Eberle, A.N., and De Graan, P.N.E. (1985) *Methods Enzymol.* **109**,129–156.

3 Brandenburg, D., Fabry, M., Schumacher, F., Strack, U. and Wedekind, F. (1990) *In:* H. Tschesche (ed.) *Modern Methods in Protein und Nucleic Acid Research*, Walter de Gruyter, Berlin, pp 305–341.

4 Wilchek, M. and Bayer, E.A. (1990) *Methods Enzymol.* **184**, 1–671.

5 Finn, F.M., Stehle, C.J. and Hofmann, K. (1985) *Biochemistry* **24**, 1960–1965.

6 Brandenburg, D., Ambrosius, D., Bala-Mohan, S., Behrendt, C., Casaretto, M., Diaconescu, C., Spoden, M., Van de Löcht-Blasberg, M. and Wedekind, F. (1988) *In*: H.J. Goren, M.D. Hollenberg and D.A.K. Roncardi (eds) *Insulin Action and Diabetes*, vol. 4, Raven Press, NY, pp 13–17.

7 Durrant, I. (1990) *Nature* **346**, 297–298.

8 Wedekind, F., Baer-Pontzen, K., Bala-Mohan, S., Choli, D., Zahn, H. and Brandenburg, D. (1989) *Biol. Chem. Hoppe-Seyler* **370**, 251–258.

9 Bodanzky, M. and Fagan, D.T. (1977) *J. Amer. Chem. Soc.* **99**, 235–239.

10 Schüttler, A. and Brandenburg, D. (1982) *Biol. Chem. Hoppe-Seyler* **363**, 317–330.

11 Fabry, M. and Brandenburg, D. (1992) *Biol. Chem. Hoppe-Seyler* **373**, 143–150.

12 Galardy, R.E., Craig, L.C., Jamieson, J.D. and Printz, M.P. (1974) *J. Biol. Chem.* **249**, 3510–3518.

13 Schüttler, A., Gattner, H.-G. and Brandenburg, D. (1984) *In*: J. Larner and S. Pohl (eds) *Methods in Diabetes Research*, Vol. 1, *Laboratory Methods*, Part A, Wiley, NY.

14 Fabry, M., Schaefer, E., Ellis, L., Kojro, E., Fahrenholz, F. and Brandenburg, D. (1992) *J. Biol. Chem.*, **267**, 8950–8956.

15 Keana, J.F. and Cai, S.X. (1990) *J. Org. Chem.* **55**, 3640–3647.

16 Bladon, C.R., Mitchell, R. and Olgier, S.S. (1989) *Tetrahedron Lett.* **30**, 1401–1404.

17 Howl, J., New, D.C. and Wheatley, M. (1992) *J. Mol. Endocrinol.* **9**, 123–129.

18 Seyer, R., and Aumelas, A. (1990) *J. Chem. Soc. Perkin Trans.* **10**, 3289–3300.

19 Ahmed, A.R.H., Olivier, G.W.J., Adams, G., Erskine, M.E., Kinsman, R.G., Branch, S.K., Moss, S.H., Notarianni, L.J. and Pouton, C.W. (1992) *Biochem. J.* **286**, 377–382.

20 Marie, J., Seyer, R., Lombard, C., Desarnaud, F., Aumelas, A., Jard, S. and Bonnafous, J.-C. (1990) *Biochemistry* **29**, 8943–8950.

21 Desarnaud, F., Marie, J., Lombard, C., Larguier, R., Seyer, R., Lorca, T., Jard, S. and Bonnafous, J.-C. (1993) *Biochem. J.* **289**, 289–297.

Immunoprecipitation of Biotinylated Cell Surface Proteins

Thomas Meier and
Hermann Leying

Summary

In order to characterize and quantitate surface or intracellular proteins, molecules are usually labeled with radioisotopes, isolated by immunoprecipitation, and analyzed by electrophoresis and autoradiography. Membrane proteins are often labeled by direct iodination; intracellular proteins are commonly labeled by growing the cells in the presence of radioactive amino acids. To avoid the problems associated with radiolabeling procedures, protocols were developed to use activated biotin for stable protein labeling followed by immunoprecipitation and sensitive detection using streptavidin-peroxidase conjugates in conjunction with chemiluminescence on Western blots. To demonstrate the reliability of these procedures, we have biotinylated extracellular domains of proteins expressed on intact human A431 cells as well as membrane proteins extracted from enriched membrane fractions and immunoprecipitated epidermal growth factor receptor (EGF-R). In combination with sensitive detection methods like chemiluminescence, this protocol offers a convenient and efficient alternative to radiolabeling of cell surfaces for the biochemical analysis of extracellular domains of membrane proteins.

5.1 Introduction

Surface antigens are commonly characterized by immunoprecipitation of radioactively labeled, detergent solubilized cell surface proteins and glycoproteins, followed by one- or two-dimensional gel electrophoresis after antibody selection under nondenaturing conditions. However, the use of high-energy radioactive labels like ^{125}I and ^{131}I isotopes has major drawbacks, such as the handling of radioactive material, limited storage time after labeling, and instability of labeled proteins due to potential radiation damage. These problems were overcome by using activated biotin for stable protein labeling, and protocols were established for combination with immunoprecipitation procedures (1, 2, 3, 4). Since biotinylated proteins are rare in nature (see Table 5.1) and a large panel of activated biotin compounds is commercially available (see Savage, this volume), this labeling technique provides a convenient alternative to radiolabeling.

Table 5.1 Biotinylated proteins in various species

Group/Species	Number of biotinylated proteins	Names of identified proteins	M_w of identified biotin-proteins (native form)	Reference
*Bacteria/*E.coli*	1	Ac-CoACO	~22 kDa	5, 6
Bacteria/others	1-3	see ref.	see ref.	6
Yeast/*S. cerevisiae*	4-5	Ac-CoACO	~190 kDa, ~250 kDa	6, 7, 5
		UreaCO	~210 kDa	
		PyrCO	~137 kDa	
Mammals/Man	4	PyrCO	~125 kDa	8, 5
		MeCro-CoACO	~75 kDa	
		Pro-CoACO	~73 kDa	
Plant	~4	Ac-CoACO	~62 kDa	9, 5
		?	~50 kDa, ~34 kDa, ~31 kDa	

Ac-CoACO: acetyl-CoA carboxylase; ureaCO: urea carboxylase; PyrCO: pyruvate carboxylase
MeCro-CoACO: methyl crotonyl-CoA carboxylase; Pro-CoACO: propionyl-CoA carboxylase.

A recent report (Au et al., 1995, *Anal. Biochem.* 226, 232) demonstrates that the patterns of endogenous biotinyl-polypeptides can be an analytical tool for bacterial identification.

Labeled proteins can be immunoprecipitated and analyzed, as before, by SDS-polyacrylamide gel electrophoresis (PAGE) (or 2-D gel electrophoresis) and detected after Western transfer using enzyme-conjugated streptavidin with a chemiluminescence substrate.

For the purpose of this review, adherent cells of the human A431 cell line were used, and biotinylation carried out with D-biotinoyl-ε-aminocapronic acid-*N*-hydroxysuccinimide ester (biotin-7-NHS) on intact cells resulting in cell surface components being specifically labeled at their extracellular domains. Additionally, biotinylation was carried out on proteins solubilized from enriched membrane fractions, by labeling intra- as well as extracellular domains of membrane components. In both cases, a monoclonal antibody to epidermal growth factor receptor (EGF-R) was used for immunoprecipitation followed by chemiluminescence detection (Fig. 5.1). The reagents used for cellular labeling and immunoprecipitation as described here are now commercially available as a kit (Boehringer Mannheim).

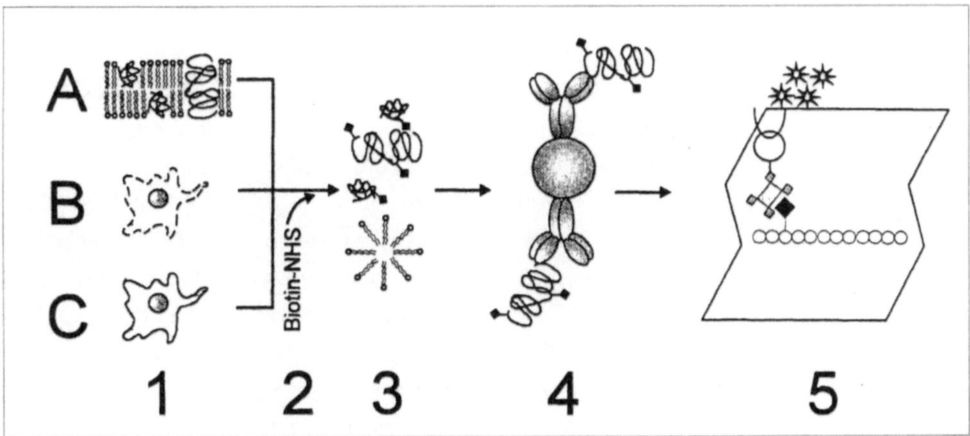

Figure 5.1 Biotin labeling of isolated membranes (1A), cell homogenates (1B), and intact cells (1C) (2, 3)

After stopping the labeling reaction, the target protein is purified by immunoprecipitation (4), separated by electrophoresis, and detected by affinity blotting using peroxidase-conjugated streptavidin and chemiluminescence substrate (5). If high-sensitivity detection is not essential, other streptavidin-enzyme conjugates and chromogenic substrates may also be used.

5.2 Technical Procedures

Cell lines

Human A431 cells derived from epidermoid carcinoma cells (10) were grown in DMEM supplemented with glucose (4.5 g/l), 2 mM L-glutamine, and 10% FCS. As a control, some experiments using enriched membrane fractions or whole-cell-labeling protocols were made with an adherent murine cell line (BALB/c-3T3 NR6) defective in EGF-R expression (11, 12). 3T3 NR6 cells were grown in DMEM supplemented with glucose (1.1 g/l), 2 mM L-glutamine, and 10% FCS.

Biotinylation of intact cells can be carried out on adherent or nonadherent cells. Nonadherent cells require minor modifications in the washing procedures to remove unreacted biotin ester (see Troubleshooting). Labeling of intact cells has the advantage that activated biotin esters specifically label the extracellular domains of cell surface proteins (2).

Protocol 5.1

Biotinylation of cell surface proteins on intact cells

1. Wash each dish 3 times with ice-cold phosphate-buffered saline (PBS) and once with borate buffer (50 mM sodium borate, 150 mM NaCl, pH 8.0).
2. Dilute D-biotinoyl-ε-aminocaproic acid *N*-hydroxysuccinimide ester (biotin-7-NHS, Boehringer Mannheim) from a stock [10 mg/ml in dimethyl sulfoxide (DMSO)] with borate buffer to a final concentration of 100 μg biotin-7-NHS/ml borate buffer. Add a total of 1.5 ml of this biotin-labeling mix to each 90-mm culture dish after completely removing borate buffer from the final washing step. Label adherent cells for 30 min at room temperature with gentle agitation.
3. To remove and inactivate remaining biotin-7-NHS, wash cells twice with excess amounts of Tris buffer (50 mM Tris, 150 mM NaCl, pH 8.3).
4. Remove cells from the dishes mechanically after incubation for 45 min in 0.02% EDTA in PBS. Wash collected cells again twice in PBS or Tris buffer.

5. Extract membrane proteins from cell pellets by sonication in extraction buffer (50 mM Tris, 150 mM NaCl, 1% NP-40, 0.5% sodium deoxycholate, 0.1 mg/ml PMSF, 1 µg/ml leupeptin, 1 µg/ml aprotinin, pH 8.0) followed by incubation for 30 min on ice.
6. Clear homogenates containing 1–2 × 10^6 cell equivalents/ml by centrifugation at 12 000 rpm on a tabletop centrifuge. Resulting supernatants can either be stored at –20 °C (avoid repetitive freeze-thaw cycles) or used directly for immunoprecipitation.

Labeling of intact cells was carried out on A431 and 3T3 NR6 cells grown to 70% confluence in 90-mm culture dishes; this is the equivalent to approximately 1–2 × 10^6 cells/dish.

Alternatively, biotinylation can also be carried out on cell membrane proteins solubilized from an enriched membrane fraction. For this protocol, an enriched membrane fraction was prepared from A431 and 3T3 NR6 cells and proteins were then detergent-extracted, followed by biotinylation with biotin-7-NHS. Under these circumstances the biotin label was not restricted to extracellular domains of cell surface proteins.

Protocol 5.2

Cell membrane preparation and biotinylation

1. Detach cells from the dish by incubation with 0.02% EDTA in PBS. Collect cells, wash twice in ice-cold PBS, and sonicate in borate buffer.
2. Clear the homogenate containing approximately 5–8 × 10^6 cells/ml by centrifugation at 12 000 rpm for 10 min on a tabletop centrifuge, and prepare an enriched membrane fraction from the supernatant by centrifugation at 100 000 × g for 1 hour at 4 °C.
3. Prior to biotin labeling, resuspend the pellet by sonication in 500 µl of 50 mM sodium borate (pH 8.0), 150 mM NaCl, 1% NP-40, 0.5% sodium deoxycholate, 0.1 mg/ml PMSF, 1 µg/ml leupeptin, 1 µg/ml aprotinin, and extract membrane proteins for 30 min on ice.
4. Biotinylate each sample with the equivalent of 100 µg of biotin-7-NHS (diluted from a 10 mg/ml stock, as above) for 15 min on ice.

5. Stop the reaction by adding NH_4Cl (from a 1 M stock) to a final concentration of 50 mM (13), and cool the sample for 20 min on ice.
6. Bring each sample to 1 ml by adding dilution buffer (50 mM Tris, 150 mM NaCl, 0.1% NP-40, 0.1 mg/ml PMSF, pH 7.5). Labeled proteins can now be immunoprecipitated using standard protocols (13).

Protocol 5.3

Immunoprecipitation

1. Incubate biotinylated protein extracts with either specific anti-EGF-R antibody or nonspecific control antibody at concentrations of 2.5 µg of antibody per milliliter of biotinylated homogenate, and incubate samples for 1–2 hours with agitation.
2. Isolate EGF-R/anti-EGF-R complexes by addition of 25 µl of protein G-agarose (Boehringer Mannheim) and incubate several hours or overnight at 4 °C.
3. For control, some samples require an initial immunoprecipitation with nonspecific IgG followed by protein G-agarose. Remove the immune complexes bound to agarose beads from the first nonspecific round, and incubate the same sample with anti-EGF-R antibody followed again by protein G-agarose incubation.
4. Wash beads twice in 50 mM Tris, 150 mM NaCl, 0.1% NP-40 (pH 7.5), followed by 2 washes in high-salt buffer (50 mM Tris, 500 mM NaCl, 0.1% NP-40, pH 7.5), and finally 1 wash in 10 mM Tris (pH 7.5) to remove salt. For this, beads can be peleted by gravity sedimentation or alternatively by centrifugation at 12 000 × g for 20–30 sec in a microfuge. Carefully remove washing solution from the bead pellet prior to the addition of 50 µl of SDS-sample buffer; boil samples if desired.

Biotinylated EGF-R was immunoprecipitated using a monoclonal antibody specific for the external domain of human EGF-R (Boehringer Mannheim). A nonspecific IgG2b antibody (Sigma) was used as negative control.

The protein G-agarose complex at this point is loaded with primary antibody and with antigen; only the latter carries the biotin label. Detection using enzyme-conjugated streptavidin will only reveal the biotinylated antigen and any labeled protein associated with the antigen (see below). In fact, recently developed protocols combining biotinylation and chemical cross-linking (see Altin, this volume) will allow specific investigation of the presence or absence of antigen-associated proteins.

For analysis of biotinylated proteins, one- or two-dimensional gel electrophoresis is appropriate. In general, the biotin label will not significantly change the molecular weight or the isoelectric point of proteins (2). With respect to high molecular weight proteins, it should be emphasized here that for Western transfer it might be appropriate to use alternative buffer compositions to the Tris/glycine/methanol system (14) most commonly used.

Protocol 5.4

SDS-PAGE, Western Transfer, and Chemiluminescence Detection

1. Transfer proteins to nitrocellulose membrane (14), rinse membrane in PBS. For the detection of low-abundant proteins it might be helpful to air-dry the membrane, since this will increase binding of proteins to nitrocellulose. Dry membranes can be stored for prolonged periods of time; rinse membranes in PBS prior to continuation.
2. Block membrane for nonspecific binding with TTBS (50 mM Tris, pH 7.4; 0.1% Tween 20; 150 mM NaCl) supplemented with 1% (w/v) blocking reagent (Boehringer Mannheim) for 1 hour with gentle agitation.
3. Incubate blot for 1–2 hours with preformed streptavidin-conjugated peroxidase (Boehringer Mannheim) at dilutions recommended by the manufacturer in the presence of blocking reagent.
4. Wash membranes for at least 1 hour in TTBS with several changes of the washing solution.
5. Detect biotinylated proteins, conjugated with streptavidin peroxidase using a commercially available chemiluminescence kit (Boehringer Mannheim) according to the manufacturer's instructions.

6. Record chemiluminescence signals on Hyperfilm ECL (Amersham) and allow prolonged exposure for weak signals to develop (15).

All samples were analyzed with SDS-PAGE using 7.5% or 5% acrylamide minigel systems (Bio-Rad) under standard conditions (13). Biotinylated molecular weight standards (Boehringer Mannheim) were used in all experiments.

5.3 Results and Discussion

Immunoprecipitation using total homogenates from cell surface biotinylated A431 cells using an EGF-R specific primary antibody and followed by chemiluminescence detection resulted in a single band with an apparent molecular weight of 160–170 kDa (Fig. 5.2, lane 1). The apparent molecular weight of this protein varied slightly depending on the percentage of acrylamide used during SDS-PAGE (data not shown). This protein band represents the fully glycosylated form of the EGF-R, which has a reported molecular weight of ~170 kDa (16, 17, 18). Control experiments, where a nonspecific IgG2b primary antibody was used during immunoprecipitation, did not result in the detection of a similar band (Fig. 5.2, lane 2). However, when the same sample was incubated with anti-EGF-R during a second round of immunoprecipitation following immunoprecipitation with nonspecific IgG2b antibody, the EGF-R specific band could be detected (Fig. 5.2, lane 3). This result, together with the finding that this band was not detected in immunoprecipitates of murine 3T3 NR6 cells which do not express the EGF-R (data not shown), demonstrates that the 170-kDa band seen in Figure 5.2 indeed represents the EGF-R.

We also compared these results of immunoprecipitation from whole-cell-biotinylated material with experiments using biotin-labeled proteins extracted from enriched membrane fractions as starting material. As was seen for the "on-dish" labeling protocol, a single 160–170 kDa protein could be immunoprecipitated under such conditions (Fig. 5.3A, lane 3). This band was not seen when the immunoprecipitation was carried out on the same extract using a nonspecific IgG2b primary antibody (Fig. 5.3A, lane 2), nor was it detectable in immunoprecipitates from murine 3T3 NR6 cells which lack EGF-receptors (Fig. 5.3A, lane 1). We therefore conclude that this single band again represents the fully glycosylated EGF-R. Moreover, prolonged chemiluminescence detection (Fig. 5.3B) resulted in the appearance of an additional ~100 kDa protein which, according to its molecular weight, probably represents the EGF-R-related protein (ERRP). This protein has a nominal molecular weight of 105 kDa and is a predominantly secreted protein related to, but not a degradation product of, the cell surface domain of the EGF-R (19).

As shown here for the EGF receptor, immunoprecipitation of biotinylated membrane proteins combined with chemiluminescence detection is a sensitive tool to study cell surface proteins without using high-energy radioisotopes. The results demonstrate that "on-dish" biotinylation of intact adherent cells is a more rapid alternative to the biotin labeling of proteins derived from enriched membrane fractions of such cells. Remaining active biotin esters can be quenched and removed easily by washing the adherent cell layer extensively with Tris-containing buffers at moderately alkaline pH. Labeled cell membrane components may be immunoprecipitated after homogenization using appropriate primary antibodies following conventional methods (13). Immunoprecipitated proteins are then analyzed by SDS-PAGE and Western transfer using protocols for the generation of chemiluminescence signals with streptavidin-peroxidase conjugates. The overall sensitivity of this method has been reported previously (2, 3).

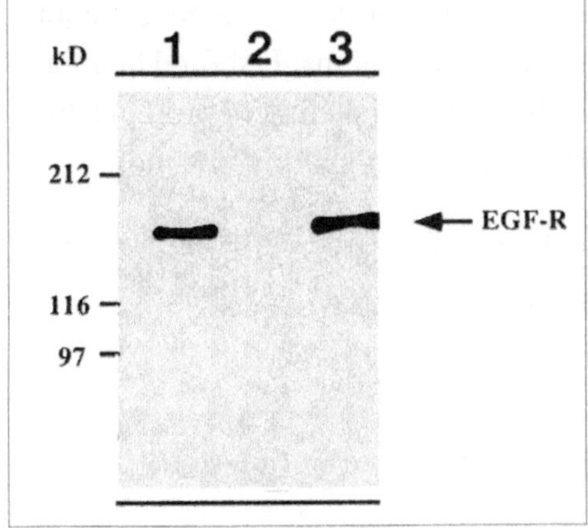

Figure 5.2 "On-dish" biotinylation of human A431 cells and immunoprecipitation of EGF-receptors

(1) A single polypeptide with the apparent molecular weight of the fully glycosylated EGF-R is immunoprecipitated by anti-EGF-R antibodies (arrow). No such protein band was detectable using a nonspecific IgG2b antibody (2). However, if the same sample was immunoprecipitated with the specific anti-EGF-R antibody during a second round of immunoprecipitation, the 160–170 kDa EGF-R protein band was detectable again (3). Positions of molecular weight markers are indicated.

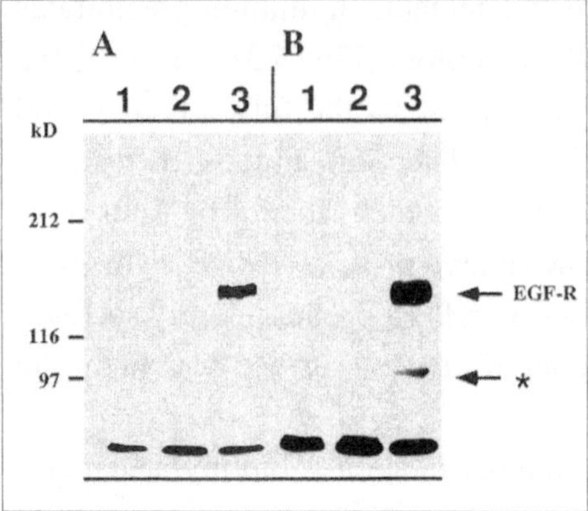

***Figure 5.3* Biotinylation of cell surface proteins solubilized from enriched membrane fractions and subsequent immunoprecipitation**

The chemiluminescence signal from the same blot was exposed to film for 2 min (A) and 30 min (B). Biotinylation of murine 3T3 NR6-membrane extracts and immunoprecipitation using anti-EGF-R antibodies (1) did not result in a specific 160–170 kDa protein band. The same was true for experiments where human A431 cell membrane fractions were extracted, biotinylated, and immunoprecipitated with nonspecific IgG2b (2). In contrast, using anti-EGF-R antibody for immunoprecipitation from biotinylated A431 cell membrane extracts resulted in the detection of a single protein band representing the human EGF-R according to its molecular weight (3). The EGF-R-related protein (ERRP) was also coimmunoprecipitated; its presence could only be detected after prolonged exposure of the chemiluminescence signal [(B), marked with asterisk]. The low molecular weight bands seen in all lanes were nonspecific, biotin-containing proteins which were also bound by the protein G-agarose. No such bands were detectable when the samples were preabsorbed with protein G-agarose prior to immunoprecipitation. Positions of molecular weight markers are indicated.

The method of protein biotinylation combined with immunoprecipitation is not restricted to cell surface components. The protocol given below may be used to biotinylate the total pool of cellular proteins after cell disruption, including recombinant proteins from eucaryotic or bacterial expression systems. It should be mentioned here that recently developed protocols allow biotin labeling of proteins during *in vitro* translation in a cell-free system (see Hoeltke et al., this volume).

Protocol 5.5

Biotin labeling of total proteins and recombinant proteins from expression systems and their preparation for immunoprecipitation

1. Wash cells at least twice with ice-cold PBS to remove any remaining serum proteins from the culture medium.

2. Add lysis buffer [50 mM sodium borate (pH 8.0) 150 mM NaCl, with 0.1 mg/ml PMSF, 1 µg/ml aprotinin and 1 µg/ml leupeptin as protease inhibitors, along with 0.5% sodium deoxycholate and 1% Triton X-100 or NP-40 as detergents] at a concentration of 10^6–10^7 cells/ml, and homogenize cells. At this point PBS or HEPES buffer could be used instead of sodium borate buffer, but Tris buffer is not compatible with the following biotinylation protocol.
3. Completely solubilize the samples.
4. Centrifuge the lysate at 12 000 x g for 10 min in the cold to remove insoluble material and transfer the supernatant to a fresh tube.
5. Add 25 µl of biotin-7-NHS stock solution (1 mg/ml) to 1 ml of the sample; mix and incubate for 15 min on ice.
6. Stop the reaction by adding 50 µl of 1 M NH_4Cl per ml sample (13) or alternatively 10 mM lysine (3), and incubate for 10–15 min on ice. The sample may then be desalted (e.g., on Sephadex G-25) to remove excess NH_4Cl and unreacted biotin ester. At this time the buffer can be changed to Tris buffer, which will additionally quench the biotinylation reaction.
7. Continue with immunoprecipitation as described in Protocol 5.3.

5.4 Troubleshooting

Sample preparation

If nonadherent cells are to be used for intact cell biotinylation, cells should be collected and washed several times in ice-cold PBS by centrifugation (13). Similarly, in order to remove unreacted biotin, cells should be washed several times in Tris buffer.

For labeling intact cells, a cell concentration not higher than 1×10^7 cells/ml should be used to ensure optimal vectorial labeling and cell recovery (2).

To reduce risk of antigen degradation during sample preparation and biotin labeling, additional protease inhibitors may be added at any step. However, the use of *N*-tosyl-L-lysine chloromethyl ketone (TLCK) and any increased concentration of peptide inhibitors containing lysine residues is not recommended, since both agents will quench the biotinylation reaction.

Best results with biotin-7-NHS are achieved in sodium borate buffer under alkaline conditions, a variety of buffers (like HEPES, PBS, carbonate, PIPES, triethanolamine) at different alkaline pH can be used for biotinylation (see 2, 20 for details). Tris buffers should be avoided at any step prior to biotinylation.

Although we never found that biotinylation of antigens subsequently interfered with their purification using specific antibodies and protein-A/G beads, this might be possible as a result of steric hindrance caused by the biotin-moiety. In this case one might try to substitute biotin-7-NHS with alternative biotin agents, labeling different target groups (see Savage, this volume for a detailed list).

In theory biotin-7-NHS is membrane permeable, although we did not encounter any problems related to this fact. Under the concentrations and incubation conditions we describe, labeling of intracellular proteins seems not to have occurred (2). If this is of concern, we suggest using sulfo-NHS-biotin, a water-soluble and membrane-impermeable agent (20).

Immunoprecipitation

According to the subclass of the primary antibody in use, either protein G-agarose, protein A-agarose, or immobilized anti-IgM antibodies should be used for immunoprecipitation. The affinity of protein A and protein G for various monoclonal antibodies is given elsewhere (e.g., ref. 13).

Samples which are likely to contain immunoglobulins (e.g., lymphoma cell lines) might need several rounds of protein A-agarose preabsorption to remove immunoglobulins, which also become biotin labeled. It is always advisable to run blank samples in which the primary antibody has been omitted. Alternatively, protein A/G-agarose can be preloaded with the desired amount of specific antibody and remaining protein A/G-binding sites can be blocked with nonspecific control antibodies or serum (13).

Nonspecific trapping of labeled proteins during washing of the protein A-agarose/antigen complexes by centrifugation may result in increased levels of background signals during detection. This can be reduced by gravity-sedimenting the complexes instead of centrifugation.

Different buffers are commonly used to wash protein A/G-antigen-antibody complexes. The tighter the binding between antibody and antigen, the more stringent the washing buffer conditions can be. The washing buffer described above is used if low-stringency conditions are appropriate. If higher stringency is needed, increase salt concentration and ionic strength by using 0.5 M NaCl or 0.5 M LiCl for the first wash. Additionally, SDS (final concentration: 0.1%) may be used during subsequent washing steps.

Western transfer, membranes, and detection

We recommend nitrocellulose over immobilon-PVDF and nylon membranes as Western blot matrix, since luminescence signal seems to be best emitted from nitrocellulose (15). The same reference provides useful tips how to optimize HRP-dependent chemiluminescence detection.

In an earlier study, we observed poor transfer using Tris-glycine-methanol transfer buffers (14), especially with high molecular weight proteins. We recommend Tris-acetate/SDS/isopropanol transfer buffers (2, 21) as an alternative.

Soaking polyacrylamide gels run under nonreducing conditions in β-mercaptoethanol (2% for 10 min) prior to protein transfer onto nitrocellulose may result in a stronger chemiluminescence signal for both disulfide- and non-disulfide-bonded proteins (22).

If high background signals during chemiluminescence detection are a problem, wash membrane once with high-salt buffers (e.g., 50 mM Tris, pH 7.5, 0.5 M NaCl, 0.05% Tween 20) for 10–15 min. However, this washing step should always be followed by a low-salt wash to decrease ionic strength prior to chemiluminescence detection.

Staining of nitrocellulose using Ponceau S is not recommended, since low-abundance antigens are unlikely to be detected, and we experienced increased levels of nonspecific background signals during luminescence detection under such conditions.

Reprobing a given blot with fresh luminol-based or chromogenic substrate is not advisable, since the peroxidase (HRP) enzyme is not stable during its reaction with the substrate and will be destroyed during the first round of exposure.

If high-sensitivity detection is not essential, chromogenic substrates for HRP-based reactions may be used instead of chemiluminescence.

Acknowledgments

The authors would like to thank Irene Huber (Boehringer Mannheim GmbH) for excellent technical support. We also thank Dr D. Hoessli and Dr S. Arni (CMU, University of Geneva) for many stimulating discussions which led to the development of the procedures described here. This work was supported in part from the Swiss National Science Foundation and by a Long Term Fellowship from the Human Science Frontier Program (T.M.).

References

1 Cole, S.R., Ashman, L.K. and Ey, P.L. (1987) *Mol. Immunol.* **24**, 699–705.

2 Meier, T., Arni, S., Malarkannan, S., Poincelet, M. and Hoessli, D. (1992) *Anal. Biochem.* **204**, 220–226.

3 Nesbitt, S.A. and Horton, M.A. (1992) *Anal. Biochem.* **206**, 267–272.

4 Altin, J.G. and Pagler, E.B. (1995) *Anal. Biochem.* **224**, 382–389.

5 Cronan, J.E. (1990) *J. Biol. Chem.* **265**, 10327–10333.

6 Fall, R.R. (1979) *Methods Enzymol.* **62**, 390–398.

7 Lim, P., Rohde, M., Morris, P.C. and Wallace, J.C. (1987) *Arch. Biochem. Biophys.* **258**, 259–264.

8 Robinson, B.H., Oei, J., Saunders, M. and Gravel, R. (1983) *J. Biol. Chem.* **258**, 6660–6664.

9 Nikolau, B.J., Wurtele, E.S. and Stumpf, P.K. (1985) *Anal. Biochem.* **149**, 448–453.

10 Fabricant, R.N., DeLarco, J.E. and Todaro, G.J. (1977) *Proc. Natl. Acad. Sci. USA* **74**, 565–560.

11 Pruss, R.M. and Herschman, H.R. (1977) *Proc. Natl. Acad. Sci. USA* **74**, 3918–3922.

12 Das, M., Miyakawa, T., Fox, C.F., Pruss, R.M., Aharonov, A. and Herschman, H.R. (1977) *Proc. Natl. Acad. Sci.* USA **74**, 2790–2794.

13 Harlow, E. and Lane, D. (1988) *Antibodies: A Laboratory Manual*, Cold Spring Harbor Laboratory Press, Cold Spring Harbor, NY.

14 Towbin H., Staehelin T. and Gordon J. (1979) *Proc. Natl. Acad. Sci. USA* **76**, 4350–4354.

15 Harper, D.R. and Murphey, G. (1991) *Anal. Biochem.* **192**, 59–63.

16 Cummings, R.D., Soderquist, A.M. and Carpenter, G. (1985) *J. Biol. Chem.* **260**, 11944–11952.

17 Gamou, S. and Shimizu, N. (1988) *J. Biochem.* **104**, 388–396.

18 Carpenter, G. and Cohen, S. (1990) *J. Biol. Chem.* **265**, 7709–7712.

19 Weber, W., Gill, G.N. and Spiess, J. (1984) *Science* **224**, 294–297.

20 Savage, D., Mattson, G., Desai, S., Nielander, G., Morgensen, S. and Conklin, E. (1992) *Avidin-Biotin Chemistry: A Handbook.* Pierce Chemical Company, Rockford, Ill.

21 Nelson, W.J. and Veshnock, P.J. (1986). *J. Cell Biol.* **103**, 1751–1765.

22 Weston, S.A., Crossett, B., Tuckwell, D.S. and Humphries, M.J. (1995) *Anal. Biochem.* **225**, 28–33.

Biotinylation and Chemical Cross-Linking of Membrane Associated Molecules

Joseph G. Altin

Summary

Cell-surface biotinylation is a popular alternative to radioiodination for cell surface labeling in studies aimed at characterizing the molecular interactions between cell surface molecules. However, such studies often require the use of chemical cross-linking agents to cross-link and preserve molecular associations upon cell disruption, and subsequent analysis of specific molecules in the cell lysate by immunoprecipitation and SDS-polyacrylamide gel electrophoresis. The procedure described below can be used to simultaneously biotinylate and chemically cross-link lymphocyte surface molecules in a single step. Briefly, the lymphocytes are washed, suspended in phosphate-buffered saline (PBS, pH 8.0), and incubated in the presence of sulfo-NHS-biotin (0.5 mg/ml) and DTSSP (0.2 mg/ml) for 30 min at room temperature. The lymphocytes are then washed and lysed in detergent; the lysates are immunoprecipitated with specific mAbs; and the immunoprecipitates are analyzed by SDS-PAGE and proteins detected by enhanced chemiluminescence (ECL). The biotinylation procedure, with or without chemical cross-linking, can also be used in conjunction with permeabilization of the cells with lysolecithin (15–25 μg/ml) to reveal associations with intracellular molecules. The advantage of performing biotinylation and chemical cross-linking on permeabilized cells, rather than in total cell lysates, is that weak molecular associations can be preserved before cell disruption. Also, washing of the permeabilized cells after treatment with reactive biotin and cross-linker avoids the need for removing these reagents from the lysate, since their presence can cause unwanted biotinylation and/or cross-linking of the antibodies used during subsequent immunoprecipitation steps. The procedure employs nonradioactive detection and is particularly useful in studies characterizing the molecular associations of cell surface receptors with other molecules, either on the cell surface or inside the cell.

6.1 Introduction

Cell-surface biotinylation has provided a convenient alternative to ^{125}I-radiodination for the labeling of cell-surface receptors and adhesion molecules (1, 2). Detection of biotinylated molecules, after SDS-polyacrylamide gel electrophoresis (PAGE) analysis and Western transfer onto nitrocellulose membrane, can be performed with high sensitivity by probing the blot with streptavidin-HRP, followed by the development of a light-producing reaction by enhanced chemiluminescence (ECL) to permanently record the information on film (3–5). The study of cell surface receptors and their interactions with ligands and other molecules on the cell surface or intracellularly, however, may require the use of chemical cross-linking agents to preserve molecular associations during subsequent analysis by immunoprecipitation and/or SDS-PAGE. Associating molecules can often be chemically cross-linked by the use of homobifunctional cross-linking reagents like dithio-bis(succinimidylpropionate)(DSP) (see, e.g., refs. 7–9), which also is available in a water-soluble form, namely, DTSSP [dithio-bis(sulfo-succinimidylpropionate)] (10). Such cross-linking reagents, however, generally have been used in conjunction with a radiolabeled ligand after surface labeling of cells by radioiodination (8–11) and/or in cells permeabilized with tetanolysin to cross-link intracellular molecules (12).

We have shown that certain lymphocyte surface molecules can be simultaneously biotinylated and chemically cross-linked in a one-step procedure, with sufficient efficiency to permit detection by nonradioactive techniques such as ECL (13). The procedure, as outlined below, can be used to label and chemically cross-link interacting cell surface receptor molecules on murine and human T lymphocytes. The procedure also can be used with lymphocytes permeabilized with lysolecithin to permit labeling and cross-linking of intracellular molecules. We see no reason why the technique cannot be applied to similar studies with other receptor molecules on other cell types.

6.2 Technical Procedures

Chemicals and monoclonal antibodies

Sulfo-NHS-biotin (sulfosuccinimidobiotin) and the reversible homobifunctional cross-linker 3,3'-dithio-bis(sulfo-succinimidylpropionate) (DTSSP) were obtained from Pierce, Rockford, IL, USA. Lysolecithin (L-α-lysophosphatidylcholine) was obtained from Sigma. The monoclonal antibodies (mAbs) used and their sources were as follows: mouse Thy-1 mAb (clone T24-31.7, ascites) and mouse CD4 mAb (clone GK-1.5, ascites) were kindly provided, respectively, by Dr I. Trowbridge, Salk Institute, La Jolla, CA, and Dr J. Ruby, Division of Cell Biology (J.C.S.M.R.); mAbs reactive with all isoforms of murine CD45 (S-450-15.2, ascites) and all isoforms of human CD45 (190-2F2.5, ascites) were a generous gift from Prof. Ian McKenzie, Austin Research Institute, Melbourne, Australia; CD3 mAb OKT3 (culture supernatant) was produced from hybridoma cells obtained from the American Type Culture Collection, Rockville, MD, USA; and rabbit polyclonal IgG Abs reactive to murine and human forms of the intracellular tyrosine kinases $p56^{lck}$, $p59^{fyn}$, and ZAP-70 were obtained from Santa Cruz Biotechnology, Santa Cruz, CA.

Cell lines and preparation of cell suspensions

The antigen-specific murine T cell clone D10 was cultured in RPMI 1640 medium (Flow Labs) supplemented with heat-inactivated 10% fetal calf serum, IL-2 (50–100 units/ml), 50 μM 2-mercaptoethanol, and antibiotics (penicillin, 120 mg/l; streptomycin, 200 mg/l; neomycin, 200 mg/l). The D10 cells were used during the rest phase and were maintained as described previously (6). The murine B lymphoma line A20 and the human leukemic T cell line Jurkat were both grown in RPMI 1640 medium supplemented with 10% heat-inactivated fetal calf serum and antibiotics (penicillin, 120 mg/l; streptomycin, 200 mg/l; neomycin, 200 mg/l). Murine thymocytes were prepared from 4 to 7 week-old female CBA mice which were bred at the John Curtin School of Medical Research. Single-cell suspensions from thymus were prepared in phosphate-buffered saline (PBS) containing 0.1% (w/v) bovine serum albumin (BSA) (Fraction V, Sigma) as described (6, 14).

Protocol 6.1

Cell surface biotinylation and chemical cross-linking

1. Wash lymphocytes 3 times with ice-cold PBS to remove contaminating fetal calf serum and other proteins from the culture medium.
2. Suspend lymphocytes at 2.5×10^7 cells/ml in PBS (pH 8.0).
3. For biotinylation only: Prepare fresh solution of sulfo-NHS-biotin (5.0 mg/ml) in PBS (pH 8.0) and mix with cell suspension (dilute 1:10 to give 0.5 mg sulfo-NHS-biotin/ml).
4. For simultaneous biotinylation and chemical cross-linking: Prepare fresh solution containing sulfo-NHS-biotin (5 mg/ml) and DTSSP (2 mg/ml) in PBS (pH 8.0) and mix with the cell suspension (dilute 1:10 to give final 0.5 mg sulfo-NHS-biotin plus 200 µg DTSSP/ml).
5. Incubate in the presence of the agent(s) for 30 min at room temperature, mix every 5–10 min.
6. Wash cells twice with cold PBS to remove unreacted sulfo-NHS-biotin (and DTSSP).
7. Solubilize cells in cold cell lysis buffer (see below), or store at –70 °C until required.

Protocol 6.2

Permeabilization of cells with lysolecithin

1. Prepare stock solution of lysolecithin 1 mg/ml in PBS containing 1% dimethyl sulfoxide (DMSO) (store frozen, thaw before use).
2. Wash lymphocytes 3 times and suspend cells at 2.5×10^7 cells/ml PBS (pH 8.0) in the presence of protease inhibitors (10 µg/ml each of leupeptin, antipain, and pepstatin, all from Sigma).
3. Add 1/40 by volume of 1 mg/ml lysolecithin (to give 25 µg/ml final concentration for Jurkat cells) and incubate for 5 min at room temperature; check cell permeability with trypan blue.
4. Add sulfo-NHS-biotin (or sulfo-NHS-biotin plus DTSSP) solution in PBS (see above) and proceed as descibed above for biotinylation with or without cross-linking.

Protocol 6.3

Preparation of detergent lysates of cells

1. Solubilize lymphocytes at 5×10^7 cells/ml by suspending cell pellet in cold lysis buffer: 1% Triton X-100 (or other appropriate detergent, e.g., 1% digitonin) in 20 mM Tris-HCl, pH 7.6, plus 150 mM NaCl (on ice).
2. Add 10% of the volume of a stock solution containing 100 mM iodoacetamide and the protease inhibitors, 100 µg/ml leupeptin, 100 µg/ml antipain, and 100 µg/ml pepstatin (all from Sigma).
3. Incubate for 30 min on ice with occasional vortexing.
4. Centrifuge for 15 min at 10 000 × g in an Eppendorf centrifuge at 4 °C to remove particulate material and nucleic acids.

Protocol 6.4

Immunoprecipitations

1. Couple specific mAbs (and control mAbs) to protein G-Sepharose beads (Pharmacia, Uppsala, Sweden) : Use 15 µl of packed beads and incubate with an aliquot of each mAb in 300 µl of PBS containing 0.5% (v/v) Triton X-100 in an Eppendorf tube on a rotator for 1–2 hours at 4 °C.
2. Wash beads twice with PBS plus 0.5% Triton X-100 to remove unbound mAb.
3. Preclear cell lysates by incubating with protein G-Sepharose for 2 hours by rotation at 4 °C: use 15 µl of packed beads to 100 µl of lysate; add 250 µl of lysis buffer to permit rotation (incubate 2 hours – overnight at 4 °C).
4. Add precleared cell lysate to Ab-coupled beads (from step 2 above) and incubate by rotation for 2 hours at 4 °C.
5. Collect immune complexes by brief centrifugation and wash either 3 times with Tris buffer (20 mM Tris-HCl, pH 7.6, 150 mM NaCl) containing 0.05% digitonin, or 3 times with PBS containing 0.5% Triton X-100, followed by 3 washes with Tris buffer.
6. To carry out re-immunoprecipitation of cross-linked proteins: Incubate washed immune complexes at room temperature in 50 µl of buffer containing 0.1% Triton X-100 and 0.5% 2-mercaptoethanol. After 5 min, add 500 µl of PBS containing 0.5%

Triton X-100, pellet the protein G-Sepharose beads, remove the supernatants, and re-immunoprecipitate the protein(s) released with appropriate Abs and wash (as above).

7. Add either reduced or nonreduced SDS-PAGE sample buffer directly to the beads, and boil for 6 min.
8. Pellet beads by brief centrifugation, and load samples in the gel slots for SDS-PAGE analysis.

Protocol 6.5

Electrophoresis, Western blotting and protein detection by enhanced chemiluminescence

1. Carry out SDS-PAGE analysis, and electrophoretically transfer proteins onto nitrocellulose membrane (Hybond-C-Super, Amersham, UK).
2. After transfer, briefly wash the membrane with PBS and air-dry at room temperature for at least 30 min (use gloves to handle membrane to avoid contamination).
3. Block the membrane by incubating in 50 ml of PBS containing 0.05% Tween 20 (PBS-Tween) and 5% low-fat skim milk.
4. Wash the membrane 4 times (~5 min each wash) with PBS-Tween.
5. Incubate the membrane with streptavidin-HRP (1:1000 in PBS-Tween, Amersham) for 60–90 min at room temperature.
6. Wash the membrane 4 times (as above) and detect proteins using the ECL protein detection system (Amersham). Immerse the membrane in a 1:1 mix of detection reagent 1 and reagent 2 (as per manufacturer's instructions).
7. Place the chemiluminescent blots inside a thin plastic bag and expose to Kodak XAR-5 film or Hyperfilm (Amersham) for recording; typical exposure times: 15 sec to 10 min.
8. Protein detection by immunoblotting can also be carried out: After protein transfer, dry the membrane and block with 5% low-fat skim milk, and probe for 1–2 hours with specific Abs, e.g., to the intracellular tyrosine kinase $p56^{lck}$. For protein detection wash the membrane and probe with protein G-HRP (1:1000, Pierce) instead of streptavidin-HRP (see above).

6.3 Results and Discussion

Biotinylation and chemical cross-linking of molecules on the lymphocyte surface

That biotinylation and chemical cross-linking of certain lymphocyte surface molecules can be carried out with sufficient efficiency to permit detection without the use of radioisotopes can be demonstrated by the association of Thy-1 and CD45 on murine T lymphocytes. Using the conventional technique of cell surface radioiodination and immunoprecipitation with appropriate Thy-1 and CD45 mAbs, an association between Thy-1 and CD45 is detected only if the cells are treated with chemical cross-linkers such as DSP to chemically cross-linking associating molecules before cell disruption (see, e.g., refs 11, 12). To study this association using nonradioactive techniques, we recently carried out experiments in which murine thymocytes were simultaneously reacted with sulfo-NHS-biotin and the reversible homobifunctional cross-linker DTSSP (13).

As shown in Figure 6.1A, nonreduced (lanes 1–5) or reduced (lanes 6–9) SDS-PAGE analysis of Thy-1 mAb T24-31.7 (lanes 2, 3, 6, and 7) and CD45 mAb S-450-15.2 (lanes 4, 5, 8, and 9) immunoprecipitates from 1% Triton X-100 lysates of murine thymocytes surface biotinylated with 0.5 mg/ml of sulfo-NHS-biotin in the presence (lanes 3, 5, 7, and 9) or absence (lanes 1, 2, 4, 6, and 8) of 200 μg/ml DTSSP. Nonreducing SDS-PAGE analysis of immunoprecipitates from lysates of surface-biotinylated thymocytes indicates that Thy-1 mAb (lane 2) and CD45 mAb (lane 4) immunoprecipitate molecules of 25–30 kDa (Thy-1) and 100–200 kDa (CD45), respectively. The smearing of the high molecular weight CD45 band (usually 170–200 kDa) is due to the heavy glycosylation of the CD45 molecule and the prolonged exposure necessary to show the fainter associated bands (see below). There were no bands in the control immunoprecipitate (lane 1). From lysates of thymocytes biotinylated in the presence of DTSSP, however, Thy-1 mAb (lane 3) immunoprecipitated the Thy-1 band (25–30 kDa) and, in addition, higher molecular weight bands of 42, 100, and 170–200 kDa, and CD45 mAb (lane 5) immunoprecipitated a broader band of 100–240 kDa, consistent with the cross-linking of these antigens to other molecules. Comparison of the Thy-1 mAb immunoprecipitate from these lysates analyzed under nonreducing (lane 3) and reducing (lane 7) conditions shows that molecules of 42, 60–65, and 180–200 kDa are co-precipitated with, and hence are re-

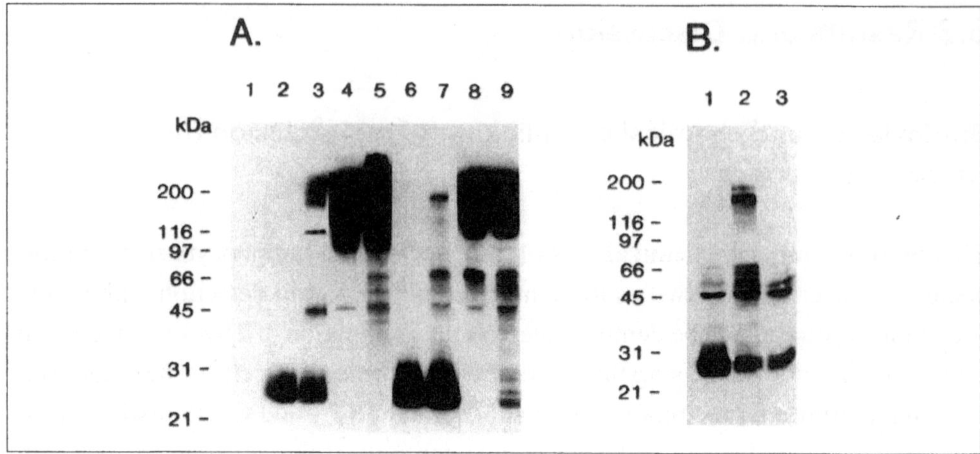

Figure 6.1. SDS-PAGE analysis of immunoprecipitates from lysates of surface biotinylated thymocytes, or thymocytes biotinylated in the presence of DTSSP

Murine thymocytes were reacted with 0.5 mg/ml sulfo-NHS-biotin in the presence or absence of 200 µg/ml DTSSP, and then washed and lysed in 1% Triton X-100. In (A) lysates from surface-biotinylated thymocytes were immunoprecipitated with control mAb (lane 1), Thy-1 mAb T24-31.7 (lanes 2 and 6), and CD45 mAb S-450-15.2 (lanes 4 and 8). Similarly, lysates from thymocytes biotinylated in the presence of DTSSP were immunoprecipitated with Thy-1 mAb (lanes 3 and 7) or with CD45 mAb (lanes 5 and 9). The immunoprecipitates were analyzed by SDS-PAGE (8–18% gradient gel) under nonreducing (lanes 1–5) or reducing (lanes 6–9) conditions. The apparent smearing or broadening of CD45 (100–220 kDa band) is due to glycosylation of the CD45 molecule (reported molecular weight 170–200 kDa) and to the prolonged exposure necessary to show up the much fainter CD45-associated bands (such as the 25–30 kDa band in lane 9). In (B) Thy-1 mAb immunoprecipitates from lysates of thymocytes biotinylated in the presence of DTSSP were eluted from the Sepharose beads by 2-mercaptoethanol cleavage of the cross-linker before subjecting the eluted proteins to re-immunoprecipitation with mAbs to Thy-1 (lane 1), CD45 (lane 2), and CD4 (GK-1.5, lane 3), and reduced SDS-PAGE analysis. The presence of the 25–30 kDa Thy-1 band in lanes 2 and 3 probably indicates leaching of the Thy-1 mAb from the original beads. The proteins were revealed by ECL detection after Western transfer onto nitrocellulose membrane and probing the membrane with streptavidin-HRP. The positions of reduced molecular weight markers are indicated in kDa. Representative results of 3 separate experiments are shown. [Reproduced from Altin and Pagler (1995) Analyt. Biochem. *224, 382–389.]*

versibly linked to, Thy-1 (lane 7). Similarly, comparison of CD45 mAb immunoprecipitates (lane 5, nonreduced; lane 9, reduced) reveals that CD45 is associated with and reversibly linked to molecules of 25–30 kDa (lane 9). The co-precipitation of the 42, 60–65, and 180–200 kDa molecules with Thy-1 (lane 7), and of the 25–30 kDa molecules with CD45 (lane 9), is not observed in immunoprecipitates from control lysates of thymocytes biotinylated in the absence of DTSSP (Thy-1 mAb, lane 6; CD45 mAb, lane 8). Thus, the fact that these molecules are observed only in immunoprecipitates from lysates of cells biotinylated in the presence of DTSSP suggests that the molecules can be reversibly cross-linked under these conditions.

The 180–200 kDa molecule seen in the Thy-1 mAb immunoprecipitate (Fig. 6.1A, lane 7) and the 25–30 kDa molecule seen in the CD45 mAb immunoprecipitate (lane 9) suggest that CD45 and Thy-1 are associated and can be chemically linked. The identity of the 180–200 kDa band in Thy-1 mAb immunoprecipitates was established by re-immunopreciptation. Lysates from thymocytes biotinylated in the presence of DTSSP were immunoprecipitated with Thy-1 mAb, and proteins were eluted from the beads by 2-mercaptoethanol cleavage of the cross-linker, before subjecting the eluted proteins in the supernatants to immunoprecipitation with CD45 mAb. As shown in Figure 6.1B, under these conditions in addition to the 25–30 kDa Thy-1 band present in all the lanes (probably due to leaching of the Thy-1 mAb from the original Sepharose beads) CD45 mAb specifically immunoprecipitated a 180–200 kDa band (lane 2) which is not seen in immunoprecipitates with Thy-1 mAb (lane 1) or CD4 mAb (GK-1.5, lane 3). This result confirms that Thy-1 mAb immunoprecipitates contain CD45, and hence that Thy-1 and CD45 can be cross-linked using this procedure.

Permeabilizing cells with lysolecithin to biotinylate and cross-link intracellular molecules

Cell surface receptors often exist in association with other cell surface receptors (15, 16) and intracellular molecules which are involved in signal transduction (see, e.g., refs 17–22). The detection of associations with intracellular molecules would, therefore, require the immunoprecipitation of the receptor with a specific mAb and then detection of any co-precipitated molecule(s), by immunoblotting with Abs to the associated molecules, or by the use of sensitive staining methods (e.g., silver or india-ink staining) to detect any co-precipitated molecule(s) in Western blots after SDS-PAGE analysis of the immunoprecipitate. Detection of the associated molecules by autoradiography after some form of metabolic labeling of the cells with radioisotopes such as ^{32}P and ^{35}S-methionine can also be used. The disadvantage of these methods, however, is that they require the use of radioisotopes or specific Abs to the unidentified molecules(s), or result in the detection of the interfering light and heavy chains of the mAb used in the immunoprecipitation step.

In order to use biotinylation to label intracellular proteins, we recently conducted biotinylation studies on cells permeabilized with lysolecithin (13), since this agent had been shown to be useful for introducing small molecules like GTP into lymphocytes (23). The biotinylation of permeabilized cells is preferable to the direct biotinylation of cell lysates, since the unreacted biotin can be removed readily from the cells by washing, thereby eliminating the potential for the unwanted labeling of the mAb used in any subsequent immunoprecipitation step. Figure 6.2 shows the permeability of the antigen-specific murine T cell clone D10, the murine B cell line A20, and the human leukemic T cell line Jurkat to the dye trypan blue after a 5-min incubation at room temperature with different concentrations of lysolecithin. The data show that treatment of D10, A20, and Jurkat cells with lysolecithin at concentrations of 15–25 μg/ml results in over 80% of the cells becoming permeable to the trypan blue (MW ~1000), and presumably also to other molecules of similar molecular size (e.g., sulfo-NHS-biotin, MW ~450).

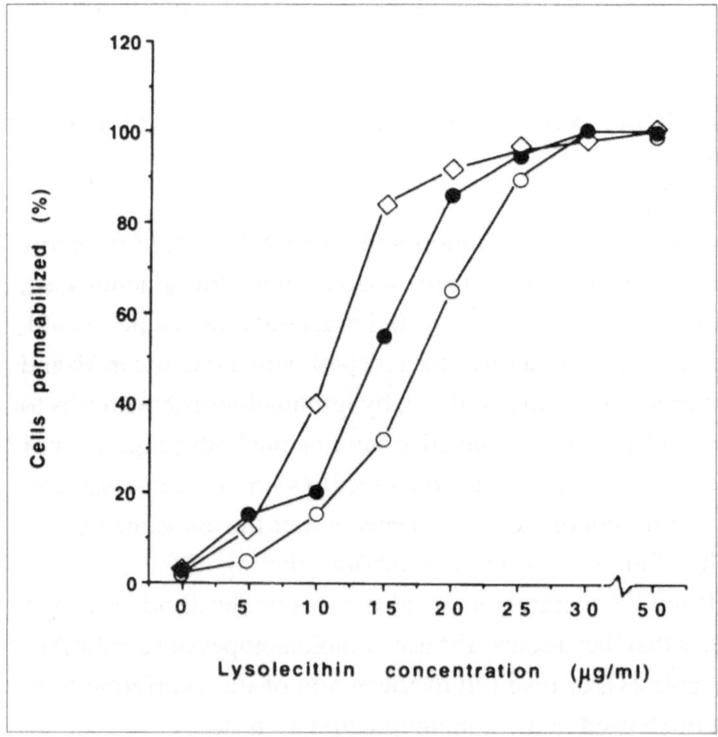

Figure 6.2 The effect of different concentrations of lysolecithin on the permeability of lymphocytes to trypan blue
Lymphocytes were washed 3 times with PBS and suspended in PBS, before incubating with the indicated concentration of lysoslecithin (0–50 μg/ml) for 5 min at room temperature, and assessing their permeability to trypan blue by counting the proportion of trypan blue-stained cells. Results for the antigen-specific murine T cell clone D10 (●), the murine B cell line A20 (◇), and the human leukemic T cell line Jurkat (○) are shown. The percentage of stained cells was only slightly (~5%) higher if assessed after 30 min of lysolecithin treatment, but depending on the cell type, complete disruption or disintegration of a proportion of cells sometimes occurred at lysolecithin concentrations >30 μg/ml (not shown). Each point represents the average of 3–5 independent determinations of the percentage of cells stained with the dye. SEM's were omitted for clarity but never exceeded ±6.5%. [Reproduced from Altin and Pagler (1995) Analyt. Biochem. *224, 382–389.]*

Jurkat cells were subjected to a 5-min pretreatment with 25 μg/ml lysolecithin, followed by treatment with sulfo-NHS-biotin, in an attempt to biotinylate both cell surface and intracellular molecules. The biotinylated cells were washed and lysed in 1% digitonin, and the lysates were used in immunoprecipitations with specific mAbs to 2 molecules functionally important in T cell activation, namely, CD3, which forms a complex with the T cell receptor (TCR) (24, 25), and CD45, whose expression is important for T cell activation (26–28). As shown in Figure 6.3A, the mAb OKT3 (Fig. 6.3A, lanes 2 and 5) specific for the ε-chain of the human CD3 complex immunoprecipitated molecules of 21–24, 38, and 45 kDa from 1% digitonin lysates of surface biotinylated Jurkat cells (lane 2); in addition, it co-precipitated molecules of 16, 36, 55–60, and 80 kDa (lane 5) from lysates of Jurkat cells biotinylated after permeabilization with lysolecithin (lane 5). Since components of the TCR:CD3 complex are reported to have molecular weights of 16 kDa (ζ-chain), 21 kDa (γ-chain), 24 kDa (ε-chain), and 38–44 kDa (TCR α- and β-chains) under reducing conditions, the observed bands of molecular weight different from these most likely represent additional molecules associated with the TCR:CD3 complex. Similarly, the mAb 190-2F2.5 to human CD45, a lymphocyte surface molecule of 170–200 kDa, immunoprecipitated a 150–200 kDa band corresponding to CD45 (Fig. 6.3A, lanes 3 and 6); co-precipitated very faint bands of 33 and 56 kDa from lysates of surface-labeled Jurkat cells (lane 3); but strongly co-precipitated molecules of 33–34, 45, 56, and 60 kDa from lysates of cells biotinylated after permeabilization (lane 6). Control immunoprecipitates from lysates of surface-biotinylated cells and cells biotinylated after permeabilization are also shown (lanes 1 and 4, respectively), and suggest that the 66-kDa band (lanes 4–6) is nonspecific.

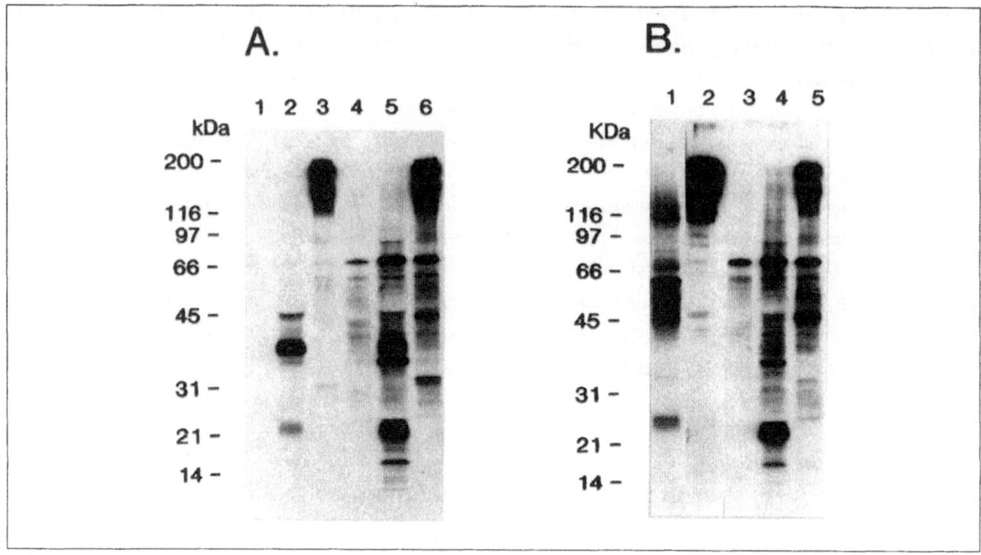

Figure 6.3. (A) SDS-PAGE analysis of CD3 and CD45 mAb immunoprecipitates from lysates of surface-biotinylated Jurkat cells, and from lysates of the cells biotinylated after permeabilization with lysolecithin

Digitonin (1%) lysates of surface-biotinylated Jurkat cells (lanes 1–3) were immunoprecipitated with control mAb (lane 1), or with mAbs to human CD3 (OKT3, lane 2) or CD45 (190-2F2.5, lane 3). Also, lysates from Jurkat cells biotinylated after permeabilization with 25 μg/ml lysolecithin (lanes 4–6) were immunoprecipitated with control mAb (lane 4), CD3 mAb (OKT3, lane 5), or CD45 mAb (190-2F2.5, lane 6). The immunoprecipitates were analyzed by SDS-PAGE (8–18% gradient gel) under reducing conditions. The reduced molecular weights of the components of the TCR:CD3 complex are reported to be 16 kDa (ζ-chain), 21 kDa (γ-chain), 24 kDa (ε-chain), and 38–44 kDa (TCR α- and β-chains); and for human CD45 the reduced molecular weight is reported to be 170–200 kDa. Therefore, the bands of molecular weights different from those of these molecules in immunoprecipitates with CD3 and CD45 mAbs most likely represent additional molecules associated with the TCR:CD3 complex and with CD45, respectively. (B) shows SDS-PAGE analysis of molecules immunoprecipitated from lysates of lysolecithin-permeabilized Jurkat cells biotinylated in the presence of DTSSP. Jurkat cells were subjected to a 5-min pretreatment with lysolecithin (25 μg/ml) and then reacted with 0.5 mg/ml sulfo-NHS-biotin in the presence of 0.2 mg/ml DTSSP before lysis in 1% Triton X-100. The lysates were immunoprecipitated with either control mAb (lane 3), CD3 mAb (OKT3, lanes 1 and 4), or with CD45 mAb (190-2F2.5, lanes 2 and 5), and analyzed by SDS-PAGE (8–18% gel) under nonreducing (lanes 1 and 2) or reducing (lanes 3–5) conditions. The proteins were revealed by ECL detection; the results are representative from 2 separate experiments. The lower intensity or absence of certain co-precipitated bands in immunoprecipitates with CD3 and CD45 mAbs in (B), compared with the corresponding immunoprecipitates in (A), presumably reflects an effect of the treatment with cross-linker in (B) and/or of the different detergent used to solubilize the cells; 1% digitonin was used in (A) and 1% Triton X-100 was used in (B). [Reproduced from Altin and Pagler (1995) Analyt. Biochem. *224, 382–389.]*

The finding that certain molecules were not detected in immunoprecipitates from lysates of surface-biotinylated Jurkat cells suggests that these molecules biotinylate poorly by cell surface labeling, or are localized intracellularly. Consistent with the intracellular localization of some of these molecules is our observation that the 56-kDa molecule in CD45 mAb immunoprecipitates from lysates of Jurkat cells (Fig. 6.3A, lanes 3 and 6) could also be detected if the same membrane was immunoblotted with Abs to the intracellular tyrosine kinase $p56^{lck}$ (not shown). This identifies the 56-kDa CD45-associated band as $p56^{lck}$, a result consistent with similar observations by other workers using other techniques (18, 29). However, immunoblotting with Abs to the $p59^{fyn}$ or the ZAP-70 tyrosine kinase also reported to associate with CD45 and the CD3 complex in some systems (20–22, 30) failed to identify the 60- and 80-kDa bands in either CD45 or CD3 mAb immunoprecipitates, suggesting that these bands may represent different molecules (data not shown).

The biotinylation of cells permeabilized with lysolecithin can also be carried out in the presence of the chemical cross-linker DTSSP. As shown in Figure 6.3B, immunoprecipitates from 1% Triton X-100 lysates of Jurkat cells treated with lysolecithin and reacted with both sulfo-NHS-biotin (0.5 mg/ml) and DTSSP (200 µg/ml) show that the CD3 mAb (OKT3) immunoprecipitated molecules of 24–26, 45–60, and 100–120 kDa when analyzed under nonreducing conditions (lane 1). Similarly, under nonreducing conditions CD45 mAb (190-2F2.5) immunoprecipitates a broad band of 120–220 kDa (lane 2). Under reducing SDS-PAGE analysis, however, both CD3 mAb (lane 4) and CD45 mAb (lane 5) immunoprecipitates contain a number of lower molecular weight bands also seen in corresponding immunoprecipitates from digitonin lysates of Jurkat cells biotinylated in the absence of DTSSP (compare Fig. 6.3B, lanes 4 and 5, with Fig. 6.3A, lanes 5 and 6, respectively). The control immunoprecipitate contains a nonspecific band at 66 kDa (lane 3). The results suggest that both cell surface and intracellular molecules can be efficiently biotinylated and cross-linked using this procedure.

These results show that cell surface and intracellular proteins can be conveniently biotinylated and cross-linked by reacting lysolecithin-permeabilized cells with sulfo-NHS-biotin and DTSSP in a one-step procedure. The use of lysolecithin to permeabilize cells is convenient and relatively inexpensive, and results in a rapid permeabilization of the lymphocytes, permitting the introduction of the reactive biotin and cross-linker into the cells; the removal of the lysolecithin is not necessary for efficient biotinylation/cross-linking of lymphocyte molecules. Thus, a 5-min pretreatment of the Jurkat cells with 25 μg/ml lysolecithin results in >85% of the cells being permeabilized. The permeabilized cells are then incubated with the lysolecithin and sulfo-NHS-biotin (with or without DTSSP) for 30 min at room temperature. By this time >90% of the cells are permeabilized, and the sulfo-NHS-biotin/DTSSP can be removed by gentle washing of the cells with PBS.

The procedure described above is simple, sensitive, and does not employ radioactive techniques. It should be noted that the usefulness of the procedure is limited to the detection of the associated molecules, as their specific identification would require immunoblotting with specific Abs after SDS-PAGE analysis and Western transfer. The procedure is particularly useful in immunoprecipitation experiments for detecting the association of specific surface receptors with other molecules either on the cell surface or intracellularly.

6.4 Troubleshooting

The concentration of lysolecithin required to yield a high proportion (>85%) of cells being permeabilized (without causing complete cell disruption) depends both on cell type and cell density. The optimal concentration of lysolecithin should be determined empirically for the particular cell type and the conditions being used. However, under the conditions decribed, the concentration of lysolecithin required for adequate labeling of intracellular molecules in lymphocytes is in the range of 15–25 μg lysolecithin/2.5 × 10^7 cells/ml for the murine B cell lymphoma A20, the murine T cell clone D10, and the human T cell line Jurkat (see Fig. 6.2).

Maximal detection of cross-linked proteins in immunoprecipitation experiments with mAbs to CD45 and to Thy-1 was achieved with DTSSP concentrations in the range 100–200 μg/ml (when the reaction was carried out for 30 min at room temperature). The use of DTSSP at concentrations above ~200 μg/ml often resulted in a reduced immunoprecipitation of the molecules by their respective mAbs, probably as a result of destruction of the antigenic site on the molecules by cross-linking. The optimal concentration of DTSSP will depend on the nature of the molecules being studied and the particular mAbs being used, and should be determined empirically for each set of conditions.

In studies where thymocytes were biotinylated in the presence of DTSSP (200 μg/ml), maximal detection of the Thy-1 and CD45 receptors, after immunoprecipitation with appropriate mAbs and analysis by SDS-PAGE, was achieved when the sulfo-NHS-biotin was used at the concentration of 0.5 mg/ml. However, the intensity of the immunoprecipitated bands was reduced several-fold if the biotinylation and cross-linking of the cells was carried out at 4 °C instead of at room temperature, or in PBS buffer of pH 7.0 instead of pH 8.0. This suggests that the biotinylation is less efficient at lower temperatures or at lower pH.

Acknowledgments

The author is greatful to Ms Eloisa Pagler for her expert technical assistance in some of the above experiments.

References

1 Hurley, W.L., Finkelstein, E. and Holst, B.D. (1985) *J. Immunolog. Meth.* **85**, 195–202.

2 Cole, S.R., Ashman, L.K. and Ey, P.L. (1987) *Mol. Immunol.* **24**, 699–705.

3 Meier, T., Arni, S., Malarkannan, S., Poincelet, M. and Hoessli, D. (1992) *Analyt. Biochem.* **204**, 220–226.

4 Nesbit, S.A. and Horton, M.A. (1992) *Analyt. Biochem.* **206**, 267–272.

5 Altin, J.G., Pagler, E.B., Kinnear, B.F. and Warren, H.S. (1994) *Immunol. Cell Biol.* **72**, 87–96.

6 Altin, J.G., Pagler, E.B. and Parish, C.R. (1994) *Eur. J. Immunol.* **24**, 450–457.

7 Hamada, H. and Tsuruo, T. (1987) *Analyt. Biochem.* **160**, 483–488.

8 Schraven, B., Samstag, Y., Altevogt, P. and Meuer, S.C. (1990) *Nature* **345**, 71–74.

9 Takeda, A., Wu, J.J. and Maizel, A.L. (1992) *J. Biol. Chem.* **267**, 16651–16659.

10 Lee, W.T. and Conrad, D.H. (1985) *J. Immunol.* **134**, 518–525 .

11 Volarevic, S., Burns, C.M., Sussman, J.J. and Ashwell, J.D. (1990) *Proc. Natl. Acad. Sci.* **87**, 7085–7089.

12 Sarosi, G.A., Thomas, P.M., Egerton, M., Phillips, A.F., Kim., K.W., Bonvini, E. and Samelson, L.E. (1992) *Intern. Immunol.* **4**, 1211–1217.

13 Altin, J.G. and Pagler, E.B. (1995) *Analyt. Biochem.* **224**, 382–389.

14 Parish, C.R., Kirov, S.M., Bowern, N. and Blanden, R.V. (1974) *Eur. J. Immunol.* **4**, 808–815.

15 Brown, M.H., Cantrell, D.A., Brattsand, G., Crumpton, M.J. and Gullberg, M. (1989) *Nature* **339**, 551–553.

16 Altevogt, P., Schreck, J., Schraven, B., Meuer, S., Schirrmacher, V. and Mitsch, A. (1990) *Intern. Immunol.* **2**, 353–360.

17 Altin, J.G., Pagler, E.B. and Parish, C.R. (1994) *Immunology* **83**, 420–429.

18 Schraven, B., Kirchgessner, H., Gaber, B.,

Samstag, Y and Meuer, S. (1991) *Eur. J. Immunol.* **21**, 2469–2477.

19 Telfer, J.C. and Rudd, C.E. (1991) *Science* **254**, 439–441.

20 Gassmann, M., Guttinger, M., Amrein, K.E. and Burn, P. (1992) *Eur. J. Immunol.* **22**, 383–386.

21 Chan, A.C., Iwashima, M., Turck, C.W. and Weiss, A. (1992) *Cell* **71**, 649–662.

22 da Silva, A.J. and Rudd, C.E. (1993) *J. Biol. Chem.* **268**, 16537–16543.

23 Peter, M.E., Hall, C., Ruhlmann, A., Sancho, J. and Terhorst, C. (1992) *EMBO J.* **11**, 933–941.

24 Altman, A., Coggeshall, K.M. and Mustelin, T. (1990) *Adv. Immunol.* **48**, 227–253.

25 Malissen, B. and Schmitt-Verhulst, A.-M. (1993) *Curr. Opin. Immunol.* **5**, 324–333.

26 Thomas, M.L. (1989) *Ann. Rev. Immunol.* **7**, 339–369.

27 Trowbridge, I.S., Ostergaard, H.L. and Johnson, P. (1991) *Biochim. Biophys. Act.* **1095**, 46–56.

28 Koretzky, G.A. (1993) *FASEB J.* **7**, 420–426.

29 Rudd, C.E., Janssen, O., Prasad, K.V.S., Raab, M., da Silva, A., Telfer, J.C. and Yamamoto, M. (1993) *Biochim. Biophy. Act.* **1155**, 239–266.

30 Samelson, L.E., Phillips, A.F., Luong, E.T. and Klausner, R.D. (1990) *Proc. Natl. Acad. Sci. (USA)* **87**, 4358–4362.

Preparation of Biotinylated Lectins and Application in Microtiter Plate Assays and Western Blotting

Elwira Lisowska, Maria Duk and Albert M. Wu

Summary

The biotinylation of lectins via amino groups with biotinamidocaproate *N*-hydroxysuccinimide ester provided reagents with fully preserved carbohydrate-binding activity which were detectable with high sensitivity using enzyme-conjugated ExtrAvidin. Lectins biotinylated with biotin hydrazide via periodate-oxidized carbohydrate residues were less sensitive reagents in avidin-mediated assays; however, the latter method may serve for identification of the lectin glycosylation status. The biotin/avidin-mediated lectin microtiter plate assay is described. Due to the high affinity of biotin-avidin interaction, this assay enables a more accurate determination of the lectin bound to the ligand-coated plate than an antibody-mediated, enzyme-linked immunosorbent assay (ELISA). The application of biotinylated lectins for detection of glycoproteins on the blot is also described and illustrated with several examples.

7.1 Introduction

Lectins are carbohydrate-binding proteins present in all kinds of organisms, from plants and microorganisms to higher animals and humans (1). The definition of lectins includes all proteins which show binding to specific carbohydrate structures and which are not anticarbohydrate antibodies or enzymes acting on carbohydrates. In the initial studies the lectins of plants and some lower animals were characterized and used mostly as tools for characterization, isolation, and fractionation of glycoproteins, glycolipids, or oligosaccharides. In recent years, the ubiquitous presence of lectins and the discovery of many biologically important processes that are dependent on lectin-carbohydrate interactions extended lectin research in the direction of studies of their biological role.

The aims of studies of the lectin-carbohydrate interaction include (i) determination of lectin specificity by measuring its interaction with structurally defined carbohydrate ligands; (ii) characterization of structures of oligosaccharide chains on glycoproteins or glycolipids using a panel of lectins with known specificity; (iii) detection of specific carbohydrate structures in cells, tissues, or other material with lectins; (iv) detection of lectins with defined oligosaccharide probes; (v) elucidation of lectin functions.

Numerous qualitative and quantitative methods are used to determine lectin-carbohydrate interactions, and most convenient are the methods in which one of the reaction partners is linked to a solid phase (cell surface, plastic surface, nitrocellulose or other sheet, thin-layer plate, etc.). In many methods, the labeled lectins or labeled secondary reagents (e.g., antilectin antibodies) are used. The label may be a radioactive isotope, fluorescent reagent, or enzyme. Fluorescent labels are most frequently used to determine lectin binding to cells in a fluorescence-activated cell sorter (FACS). Radiolabeling gives a high sensitivity of the assay, but radioisotopes have limited stability and create a health hazard. For these reasons, radioactive methods have recently been replaced by enzymatic methods. A popular version of such a method is the enzyme-linked assays in which antilectin antibodies and enzyme-linked secondary anti-Ig antibodies are used. However, this method requires obtaining antibodies against any lectin studied, which may be laborious, especially if a great number of lectins is assayed. This difficulty can be overcome by the use of another enzymatic method in which biotinylated lectins are detected with enzyme-linked avidin. This sensitive and versatile method has frequently been used to detect lectins bound to tissue sections or to glycoproteins immobilized on Western blots (2–4) and has occasionally been applied in microtiter plate assays. In most studies commercially available biotinylated lectins are used. However, biotinylation of lectins is a simple procedure that can easily be performed in the laboratory (5, 6) and is especially useful when many lectins in untreated and biotinylated form are to be studied. In this chapter we describe the biotinylation of lectins and application of biotinylated lectins in microtiter plate assays and in overlay assays on Western blots.

7.2 Technical Procedures

Materials

The lectins were commercial preparations purchased from Sigma (St. Louis, MO, USA). Biotin hydrazide, biotinamidocaproate-*N*-hydroxysuccinimide ester, ExtrAvidin-alkaline phosphatase (Extravidin-AP), ExtrAvidin-horseradish peroxidase (ExtrAvidin-HRP), enzyme substrates (Sigma 104 phosphatase substrate 5-mg tablets and 4-chloro-1-naphthol) were also purchased from Sigma. Ninety-six-well flat-bottom microtiter ELISA plates (Nunc, MaxiSorp, Vienna, Austria) and nitrocellulose BA85 (Schleicher & Schuell, Dassel, Germany) were used. Glycophorin A (GPA) was purified from human erythrocyte membranes (7). Asialoglycophorin A (asGPA) and asialo-agalactoglycophorin A (asagGPA) were obtained as described previously (5).

Buffers

PBS – 10 mM phosphate buffer in 0.15 M NaCl, pH 7.3
TBS – 50 mM Tris-HCl buffer in 0.15 M NaCl, pH 7.3
TBS-T – TBS containing 0.05% Tween 20
Coating buffer – 50 mM carbonate buffer, pH 9.6
Substrate buffer for alkaline phosphatase – 50 mM carbonate buffer of pH 9.6 containing 1 mM $MgCl_2$

Biotinylation of lectins

The lectins were biotinylated using 2 methods, biotinamidocaproate-*N*-hydroxy-succinimide ester (called biotin ester) via lectin amino groups, and biotin hydrazide via lectin-oxidized carbohydrates (Fig. 7.1). The protocol of biotinylation follows.

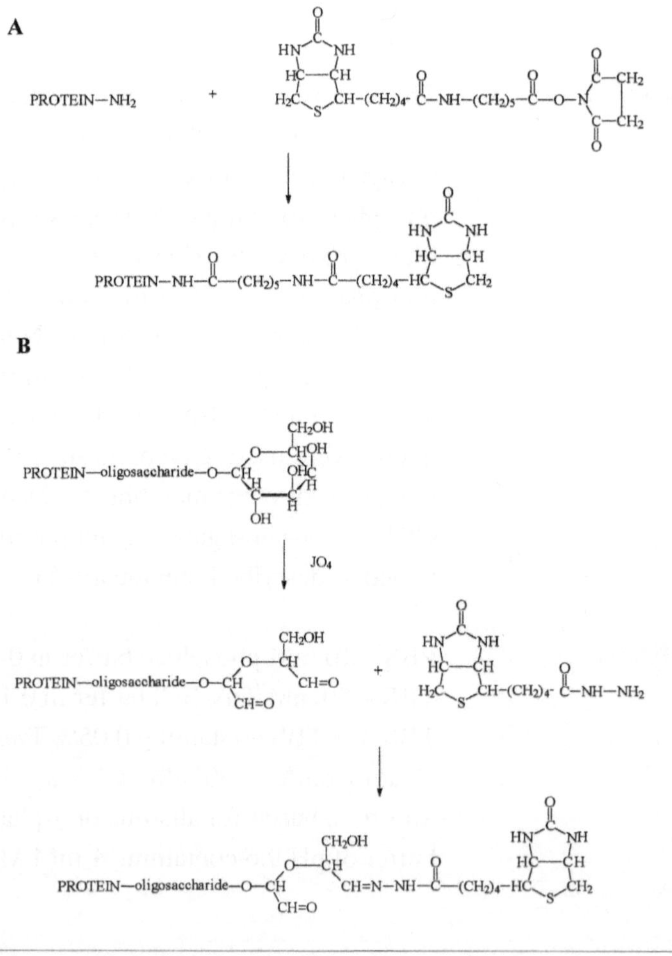

A

PROTEIN—NH_2 + biotinamidocaproate-N-hydroxysuccinimide ester → PROTEIN—NH—C(=O)—$(CH_2)_5$—NH—C(=O)—$(CH_2)_4$—biotin

B

PROTEIN—oligosaccharide—O—(sugar ring) —JO_4→ PROTEIN—oligosaccharide—O—CH(CH=O)—O—CH(CH_2OH)—CH=O + biotin—$(CH_2)_4$—C(=O)—NH—NH_2 → PROTEIN—oligosaccharide—O—CH(CH=O)—O—CH(CH_2OH)—CH=N—NH—C(=O)—$(CH_2)_4$—biotin

Figure 7.1 The scheme of biotinylation reactions with biotinamido-caproate-N-hydroxy-succinimide ester (A) and with biotin hydrazide (B)

Protocol 7.1 **Lectin biotinylation**

A. Biotinylation with biotinamidocaproate-N-hydroxy-succinimide ester

1. Prepare the solution of 200 μg of lectin in 250 μl of PBS.
2. Dissolve 500 μg of biotin ester in 50 μl of methanol, and mix with 1.95 ml of PBS.
3. Mix 250 μl of lectin solution with 400 μl of biotin ester solution and incubate for 30 min at room temperature.
4. Dialyze the sample at 4 °C for 2–3 hours against water and then overnight against PBS or TBS.
5. Complete the sample of biotinylated lectin with 5 μl of 20% sodium azide and with PBS or TBS to a final volume of 1 ml. This sample is ready to use (after proper dilution) and can be stored in the refrigerator for several weeks.

B. Biotinylation with biotin hydrazide

1. Prepare the solution of 200 μg of lectin in 180 μl of distilled water, cool to 0 °C on ice-water bath.
2. Prepare 0.2 M sodium periodate in water, cool to 0 °C.
3. Add 20 μl of sodium periodate to the lectin sample and incubate for 1 hour at 0 °C.
4. Prepare the solution of 0.025% biotin hydrazide in PBS.
5. Treat the oxidized lectin sample with 10 μl of ethylene glycol, mix and incubate for 10 min at room temperature.
6. Add 400 μl of biotin hydrazide solution, mix and incubate for 30 min at room temperature.
7. Further treatment as in A, steps 4 and 5.

The microtiter plate assay

The binding of biotinylated lectins to glycoprotein-coated ELISA plates was tested. The lectins bound were detected with ExtrAvidin-alkaline phosphatase conjugate using appropriate substrates (Sigma 104 phosphatase substrate tablets), as described in Protocol 7.2. The approximate concentrations of the coating substance and biotinylated lectins we used are given, but it should be stressed that optimal concentrations for different lectins and lectin ligands may vary and should be selected experimentally, taking into account the aim of the assay. The assay should be performed in duplicate (in our experiments standard deviation did not exceed 5%) and should include the following controls: (i) wells coated with buffer only (glycoprotein omitted); (ii) wells in which biotinylated lectin is omitted (buffer added). In our experience the absorbance values for control wells were below 0.05, which indicated that in the presence of Tween 20 used in the assay an unspecific binding was negligible.

Protocol 7.2

Microtiter plate lectin assay

1. Fill the wells of ELISA plate with glycoprotein solution in 0.05 M carbonate buffer, pH 9.6 (2–20 µg/ml, 50 µl per well), cover with plastic or with Parafilm and leave overnight at 4 °C, or for 2 hours at 37 °C.
2. Wash the plate 3 times with TBS-T (200 µl/well).
3. Add biotinylated lectin solution in TBS-T (50 µl/well, earlier selected lectin concentration or serial dilutions of the lectin (e.g., in the range of 0.01–10 µg/ml = 0.5–500 ng/well) and leave for 1 hour at room temperature.
4. Wash the plate as in step 2.
5. Add ExtrAvidin-AP solution diluted 1:5 000 with TBS-T and incubate for 1 hour at room temperature.
6. Wash the plate 5–6 times with with TBS-T (200 µl/well).
7. Dissolve Sigma 104 phosphatase substrate 5-mg tablet in 5 ml of 0.05 M carbonate buffer, pH 9.6, containing 1 mM $MgCl_2$.
8. Add to each well 50 µl of the substrate solution.
9. Read the absorbance at 405 nm in the microtiter plate reader as the color develops, usually after 30–60 min.
10. The enzymatic reaction can be stopped by adding 25 µl/well of 0.3 M NaOH.

Lectinoblotting

The glycoproteins or cell extracts were fractionated by SDS-PAGE in 10% polyacrylamide gel according to the procedure of Laemmli (8) and were electrophoretically transferred from the gels to nitrocellulose sheets (9). The protocol of detection of glycoprotein bands with biotinylated lectins is presented below. The reactivity of a lectin with the material studied (e.g., cell extract, blood serum, or purified glycoprotein) can be checked before electrophoresis by the dot-blot method. For this purpose, 1–10 μg of protein is applied to a nitrocellulose sheet in the form of a dot and the sheet is treated in the same way. When a biotinylated lectin is submitted to the electrophoresis and blotting, the blot is treated as described in Protocol 7.3, except that steps 3 and 4 are omitted.

Protocol 7.3

Glycoprotein detection with biotinylated lectins in Western blotting

1. Overlay the blot with transferred glycoproteins with 5% BSA in TBS and incubate for 1 hour at 37 °C or overnight at 4 °C.
2. Wash the blot with 0.15 M NaCl in water or with PBS, 2 times for 5 min with gentle shaking.
3. Overlay the blot with biotinylated lectin solution in TBS-T (1–5 μg/ml) and incubate for 1 hour at room temperature with gentle shaking.
4. Wash with 0.15 M NaCl, 5 times for 10 min.
5. Overlay the blot with ExtrAvidin-HRP conjugate (diluted 1:2 000 with TBS-T) and incubate for 1 hour at room temperature with gentle shaking.
6. Wash, as in step 4.
7. Prepare the HRP substrate solution: 6 μg of 4-chloro-1-naphthol dissolved in 2 ml of methanol; add 10 ml of TBS and 6 μl of 30% H_2O_2.
8. Immerse the blot in the substrate solution and wait until the color develops (usually a few minutes).
9. Stop the enzymatic reaction by transferring the blot from the substrate solution to water.

7.3 Results and Discussion

The most popular method of protein biotinylation is binding of "active" biotin esters to protein amino groups (10). Since most lectins are glycoproteins, we have also tried biotinylation by linking biotin hydrazide to aldehyde groups generated in carbohydrates by periodate oxidation. We compared both methods of biotinylation on a panel of lectins, checking the effect of modification on their agglutinating activity. Biotinylation with biotin ester does not decrease lectin activity; and while biotinylation with biotin hydrazide does decrease the binding activity of some lectins, this effect is caused by periodate oxidation, not by biotinylation (Table 7.1, DBA, LBA, WFA, UEA-I). Moreover, all lectins biotinylated with hydrazide (including those which were fully active) were less sensitive reagents in the microtiter plate assay than were lectins modified with biotin ester (examples are shown in Fig. 7.2). Nevertheless, modification with biotin hydrazide may be the method of choice when blocking lectin amino groups is undesirable. Moreover, this method can be applied to test whether the lectin is glycosylated. To this aim, SDS-PAGE of biotinylated lectin, blotting, and detection of the lectin band with enzyme-conjugated avidin are performed. As shown in Figure 7.3, periodate/biotin hydrazide-treat-

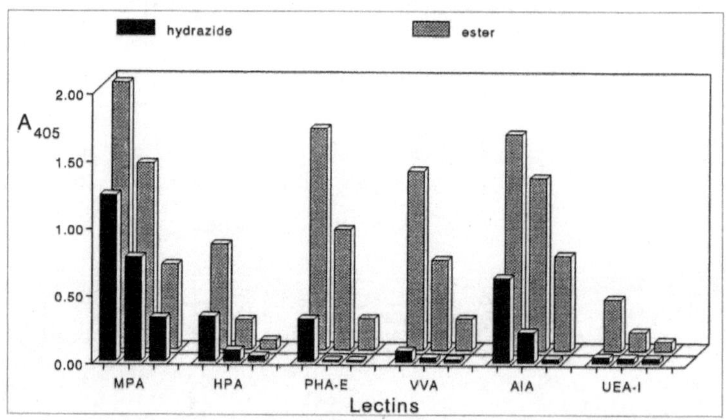

Figure 7.2 ExtrAvidin detection of biotin hydrazide- and biotin ester-labeled lectins linked to the coated ELISA plate

The plate was coated with asGPA (0.2 µg/well), except for tests with VVA and AIA, where the plate was coated with asagGPA (0.2 µg/well). Each lectin was used at the concentrations of 1, 0.25, and 0.06 µg/ml (triplicate bars from left to right) and absorbance at 405 nm is given. VVA, Vicia villosa agglutinin B4; for abbreviations of other lectins see Table 7.1.

Table 7.1 Agglutinating activity of biotinylated lectins

Lectin source (abbreviaton)	**Specificity**	**Erythrocytes**	**Concentration required for agglutination (μg/ml)**			
			untreated	oxidized	hydrazide	*ester**
Dolichos biflorus	αGalNAc	A1	2.5	*20.0*	*20.0*	2.5
(DBA)		asialoA_1	0.3	*1.2*	*1.2*	2.5
Helix pomatia	GalNAc	A1	0.3	0.3	0.3	0.3
(HPA)		asialoA_1	0.08	0.1	0.1	0.08
Phaseolus lunatus	GalNAc	A1	5.0	*>20.0*	*>20.0*	5.0
(Lima bean, LBA)		asialoA_1	0.3	*>10.0*	*>10.0*	0.3
Glycine max	GalNAc	A1	2.5	2.5	2.5	2.5
(soya bean, SBA)		asialoA_1	0.2	0.3	0.3	0.2
Wistaria floribunda	GalNAc	A1	2.5	5.0	5.0	2.5
(WFA)		asialoA_1	0.4	0.8	0.8	0.4
Ulex europeus	Fuca1-	0	1.2	*5.0*	*5.0*	1.2
(UEA-I)	2Gal	asialo0	0.3	*1.5*	*1.5*	0.3
Maclura pomifera (MPA)	GalNAc	asialo0	0.02	0.02	0.02	0.02
Phaseolus vulgaris (PHA-E)	bisected *N*-glycans	asialo0	0.2	0.3	0.3	0.2
Lens culinaris (LCA)	Man/Glc	asialo0	1.2	1.2	1.2	1.2
Ricinus communis (RCA)	Gal	asialo0	0.08	0.08	0.08	0.08
Artocarpus integrifolia (AIA)	GalNAc	asialo0	0.04	0.04	0.04	0.04

* Periodate-oxidized lectin modified with biotin hydrazide

** Lectin biotinylated with biotin ester

Comparison of periodate oxidation combined with biotin hydrazide reaction and biotinylation with biotinamidocaproate-N-hydroxysuccinimide ester.

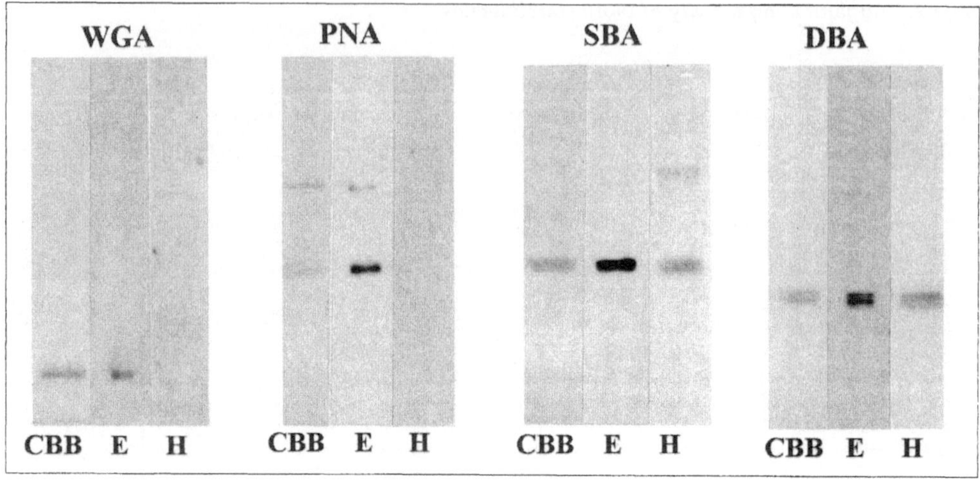

Figure 7.3 Detection of biotinylated lectins after SDS-PAGE and blotting with ExtrAvidin-HRP
CBB, the gels stained with Coomassie Brilliant Blue, 4 µg of protein per lane; E and H, lectins modified with biotin ester or biotin hydrazide, respectively, transferred from polyacrylamide gels to nitrocellulose and detected with ExtrAvidin-HRP; E, 0.2 µg of lectin protein per lane, ExtrAvidin diluted 1:2000; H, 4 µg of lectin protein per lane, ExtrAvidin diluted 1:1000.

ed peanut agglutinin (PNA) and wheat germ agglutinin (WGA), which are not glycosylated, were not detected with avidin, whereas similarly treated glycosylated lectins (DBA and SBA) gave distinct bands.

All further experiments were performed with lectins biotinylated via amino groups.

As a ligand for testing the binding of biotinylated lectins in the plate assay we used glycophorin A (GPA) from human erythrocytes (11). GPA contains disialylated Galβ1-3GalNAc-*O*-linked chains, which can be easily desialylated by a mild acid hydrolysis. In turn, Gal residues can be removed from asialoGPA (asGPA) by Smith degradation (periodate oxidation followed by a mild acid hydrolysis). These modifications, which can be performed not only in solution but also on glycoproteins immobilized on ELISA plates or nitrocellulose (5, 6), make GPA a convenient ligand for anti-T (Galβ1-3GalNAc-Ser/Thr) and anti-Tn (GalNAc-Ser/Thr) lectins. Moreover, GPA has 1 *N*-linked biantennary chain of complex type with the bisecting GlcNAc residue and binds several *N*-glycan-specific lectins. The enzyme-conjugated avidin is used for detection of the bound lectin, and usually hen egg avidin or bacterial streptavidin is used. The first is more efficient in biotin binding but is glycosylated, and this may produce side reactions (12). It is particularly important in lectin studies, when crude biological material (which may contain endogenous lectins) is used, or when competitive experiments using biotinylated and unlabeled lectins are performed. Use of streptavidin avoids these side effects, but streptavidin has a lower biotin-binding capacity. In our experiments we used enzyme-conjugated ExtrAvidin, characterized by the manufacturer (Sigma) as a reagent combining the high specific activity of avidin and the low background binding of streptavidin.

The biotin/avidin-mediated microtiter plate assay shows generally similar sensitivity as immunoenzymatic assays (5, 13). To compare both methods directly, binding of biotinylated jacalin (AIA) and *Vicia villosa* agglutinin (VVA) to an asialoagalactoGPA (asagGPA)-coated plate was measured by means of enzyme-linked ExtrAvidin or with antilectin rabbit antibodies (Ab) and goat antirabbit Ig enzyme-linked Ab. The preparation of antilectin antibodies and the immunoenzymatic assay were performed as described earlier (13). The ELISA plates were coated with asagGPA at a concentration of 4 µg/ml (0.2 µg/well), which gave the highest binding of most lectins (5). There was no difference between untreated and biotinylated lectins in the Ab-mediated assay (not shown). However, the binding curves at the range of lectin concentration of 4–1000 ng/ml (0.2–50 ng/well) were different in immunoenzymatic and biotin/avidin-mediated assays (Fig. 7.4A). In the Ab assay the lectins showed apparent saturation of the plate at the highest lectin concentration used. The avidin assay, which seems to be less sensitive at low lectin concentrations, showed higher and strongly increasing binding of both lectins at their higher concentrations, and saturation of wells was not reached. The same experiment was repeated with the plate coated with asagGPA at a 160-fold-lower concentration (25 ng/ml) and with a broader range of lectin concentrations, to achieve saturation of wells in the avidin assay. Binding of lectins was still detectable under these conditions by both methods (Fig. 7.4B). VVA showed lower binding than AIA, and its binding curves obtained with antibodies or avidin had a similar shape, indicating that the system had reached saturation. The binding curve of AIA obtained with avidin was steeper than the flat curve obtained with antibodies. These results suggested that in the Ab assay a limiting factor "flattening" the binding curves is an insufficient excess of one or both antibodies used. This effect depends on the concentration and affinity of the antibodies. In our system an increase of the antilectin Ab concentration did not change the results, but the use of a 4-times-lower dilution of the second Ab (1:250 v. 1:1000, a dilution 1:1500–1:4000 was recommended by the producer) for testing AIA gave almost 2-fold-higher absorption values (not shown). However, a desirable increase of the second Ab concentration is not convenient because it is connected with higher background absorption values and higher cost of the reagent. Due to its high affinity for biotin (several orders of magnitude higher than average Ab affinity) avidin allows more accurate measurement of biotinylated lectin.

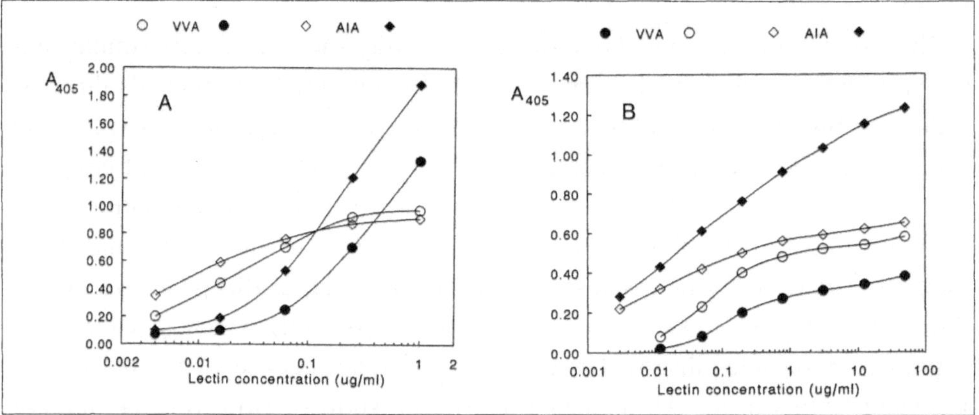

Figure 7.4 Comparison of biotin/avidin- and antibody-mediated microtiter plate assays with the use of VVA and AIA
The biotin/avidin assay (black symbols) was performed according to Protocol 7.2. In the antibody assay (open symbols) the lectins were detected with rabbit anti-VVA and anti-AIA antibodies and with AP-conjugated anti-rabbit Ig antibodies (Dako, Denmark) diluted 1:1000 (13). The plate was coated with asagGPA at a concentration of (A) 4 μg/ml (0.2 μg/well) or (B) 25 ng/ml (1.2 ng/well). The data points represent the mean of duplicate

Besides testing the binding or inhibition of lectin binding with various high or low molecular weight ligands (5), biotinylated lectins can be used in other versions of the plate assay. For example, competitive tests can be performed in which the binding of biotinylated lectin is inhibited by an excess (or serial dilutions) of unlabeled lectin. This method can be used for comparison of relative affinities of lectins specific for the same carbohydrate structure or for comparison of the specificity of lectins. In inhibition assays it is necessary to determine the complete binding curve to select the proper concentration of biotinylated lectin (usually a point in the upper part of the slope of the binding curve). Complexing biotinylated lectin with its ligand in solution (e.g., cell extract, body fluid) and binding of the complex to an avidin-coated surface can be used as an analog of immunoprecipitation assays (14).

Biotinylated lectins are widely used for identification of glycoprotein bands in blots. The method is sensitive (see Protocol 7.3), and gives distinct glycoprotein bands and low background staining. A few examples of lectinoblotting are shown in Figure 7.5. It is noteworthy that only 6 µg of total erythrocyte membrane proteins per lane is sufficient for distinct detection of all glycophorin bands with PNA on the desialylated blot, and 1 µg of weakly glycosylated ovoalbumin (containing a small percentage of sugars only) is sufficient for its detection with ConA. This procedure is used to characterize the glycosylation profile in biological material, identify protein glycosylation, and characterize the structure of oligosaccharide chains with a panel of lectins. The possibility of chemical modification of glycoprotein oligosaccharide chains on the blot (Fig. 7.5) enables the use of lectins that do not bind to the native structures but to the internal structures of sugar chains (6).

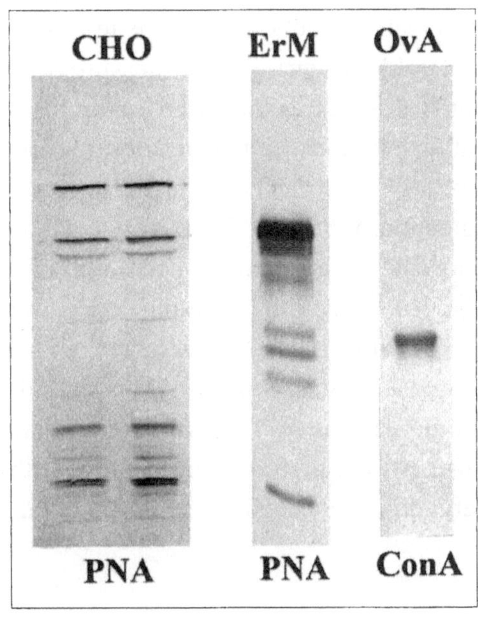

Figure 7.5 Application of biotinylated lectins for detection of glycoproteins separated by SDS-PAGE and transferred to the nitrocellulose blots (according to Protocol 7.3)
CHO, the NP40 extracts of 2 lines (Pro5 and C5) of Chinese hamster ovary (CHO) cells (15), 40 µg of protein per lane probed with the biotinylated peanut agglutinin (PNA) at a concentration of 5 µg/ml; ErM, the SDS lysate of erythrocyte membranes obtained according to the method of Dodge et al. (16), 6 µg of protein per lane, the blot was desialylated by incubation in 0.025 M sulfuric acid for 1 hour at 80 °C, the biotinylated PNA (2.5 µ/ml) gives staining of all monomeric and dimeric forms of glycophorins A, B, and C; OvA, purified ovoalbumin, 1 µg per lane, staining by means of biotinylated concanavalin A (ConA), 3 µg/ml.

7.4 Troubleshooting

Methodological problems connected with the use of biotinylated lectins are not confined to the biotin/avidin system used but rather to the methods in which this system is applied. Concerning lectin biotinylation, it is important to check that this modification does not decrease the carbohydrate-binding activity of the lectin and that the lectin-bound biotin residues are detected with avidin with sufficient sensitivity. In such methods as lectinoblotting or plate lectin assay the same problems may arise, and the same rules must be followed as in immunoblotting and plate ELISA, respectively. Avidin conjugated with HRP or AP can be used in both methods. However, HRP gives faster and more intense color development than AP in the plate assay and is more suitable for qualitative assays (lectinoblotting, or checking the positive reaction on the plate). On the other hand, the avidin-AP conjugate is preferable for quantitative measurements on the plate because slower color development decreases the "time" error, the plate can be read several times after different time periods, and the results are more accurate and reproducible. The entire detection system can be changed to more sensitive methods like chemiluminescence detection. We found that binding of biotinylated lectins to wells is not always proportional to the ligand concentration in the coating solution and may be lower at higher-than-optimal "coating concentrations" (5). Beginning the work with a new glycoconjugate/lectin system, the binding of serially diluted lectin samples to wells coated at various concentrations should be determined, and the desired conditions should be selected for further (e.g., inhibition) studies. When comparative inhibition tests are performed in parallel at the same lectin and coating substance concentrations, conclusive results are obtained.

Acknowledgments

This work was supported by Grants from the Chang-Gung Medical Research Project (CMRP No. 293), Kwei-.san, Tao-yuan, and the National Science Council (NSC 84-2331-B-182-016 & 85-2331-B-182-079), Taipei, Taiwan.

References

1 Sharon, N. and Lis, H. (1989) Lectins. Chapman and Hall, London.
2 Wilchek, M. and Bayer, E.A. (1990) *Methods Enzymol.* **184**, 14–45.
3 Hsu, S.M. (1990) *Methods Enzymol.* **184**, 357–363.
4 Pino, R.M. (1990) *Methods Enzymol.* **184**, 388–395.
5 Duk, M., Lisowska, E., Wu, J.H. and Wu, A.M. (1994) *Anal. Biochem.* **221**, 266–272.
6 Wu, A.M., Duk, M., Lin, M., Broadberry, R.E. and Lisowska, E. (1995) *Transfusion* **35**, 571–576.
7 Lisowska, E., Messeter, L., Duk, M., Czerwiñski, M. and Lundblad, A. (1987) *Molec. Immunol.* **24**, 605–613.
8 Laemmli, U.K. (1970) Nature **227**, 680–685.
9 Towbin, H., Staehelin, T. and Gordon, J. (1979) *Proc. Natl. Acad. Sci. USA* **76**, 4350–4354.
10 Bayer, A.E. and Wilchek, M. (1990) *Methods Enzymol.* **184**, 138–160.
11 Lisowska E. (1988) Antigenic Properties of Human Erythrocyte Glycophorins. In: Wu, A.M. and Adams, L.G. (eds) *Molecular Immunology of Complex Carbohydrates*, Plenum Press, New York, pp 265–315.
12 Duhamel, R.C. and Whitehead, J.S. (1990). *Methods Enzymol.* **184**, 201–207.
13 Duk, M., Mitra, D., Lisowska, E., Kabat, E.A., Sharon, N. and Lis, H. (1992). *Carbohydr. Res.* **236**, 244–258.
14 Cook, G.M.W. and Buckie, J.W. (1990) *Methods Enzymol.* **184**, 304–314.
15 Remaley, A.T., Ugorski, M., Wu, N., Litzky, L., Burger, S.R., Moore, J.S., Fukuda, M. and Spitalnik, S.L. (1991) *J. Biol. Chem.* **266**, 24176–24183.
16 Dodge, J.T., Mitchell, C. and Hanahan D.J. (1963) *Arch. Biochem. Biophys.* **100**, 119–130.

Biotin-Labeling of Poly(ADP-ribose) in Poly(ADP-ribose)-Protein Interactions

Frank M. Narendja
and Georg Sauermann

Summary

Methods are described for the preparation of poly(ADP-ribose) and for the preparation of biotinylated poly(ADP-ribose). Poly(ADP-ribose) is covalently modified to its biotinylated derivative by light-induced reaction of photobiotin with the polymer. Furthermore, conditions are given for the detection of poly(ADP-ribose) binding proteins in ligand-blotting experiments. The electrophoretically separated proteins are identified on blots using biotinylated poly(ADP-ribose) in a staining reaction. The results obtained with biotinylated poly(ADP-ribose) are comparable with those obtained with radioactive poly(ADP-ribose).

8.1 Introduction

Poly(ADP-ribosyl)ation is one of the posttranslational modifications of nuclear proteins. The nuclear enzyme poly(ADP-ribose) polymerase (EC 2.4.2.30) catalyzes the formation of poly(ADP-ribose) chains covalently linked to acceptor proteins. The activity of the DNA-dependent enzyme is greatly stimulated by DNA strand breaks. Poly(ADP-ribosyl)ation reactions have been related to the regulation of several cellular functions, e.g., DNA replication, recombination, and repair. However, neither the main biological role nor the biochemical mechanism of action of ADP-ribosylation reactions has been unequivocally established. Many interpretations proceed from an alteration of protein properties as induced by covalent protein modification (for reviews, see refs 1–3).

On the other hand, the potential role of poly(ADP-ribose) chains *per se* on nuclear functions has also been investigated (4). Using a filter-binding assay, the capability of poly(ADP-ribose) to affect DNA-histone interaction was demonstrated (4). Nuclear nonhistone poly(ADP-ribose) binding proteins were found (5, 6) via ligand blotting.

In most published studies poly(ADP-ribose) was detected by radiolabeling techniques. In other fields, however, as in investigations of DNA-protein and RNA-protein interactions, substitution of radiolabeled probes by biotin-modified probes has proved advantageous. We therefore investigated whether biotinylated poly(ADP-ribose) could be conveniently prepared and usefully applied in studies on poly(ADP-ribose)-protein interactions (6).

In this chapter we present the experimental procedure for the biotinylation of poly(ADP-ribose) using photobiotin, a reagent developed by Forster et al. (7). The covalent modification is mediated by light-induced reaction of the aryl-azido group of photobiotin with poly(ADP-ribose). Furthermore, the application of biotinylated poly(ADP-ribose) in ligand-blotting experiments is described.

8.2 Technical Procedures

Purification of poly(ADP-ribose) polymerase and preparation of poly(ADP-ribose)

Materials and buffers
(Protocol 8.1 and 8.2)

- Fresh calf thymus
- Blender
- Extraction buffer: 50 mM Tris-HCl, pH 8.0, 300 mM NaCl, 10 mM 2-mercaptoethanol, 50 mM $NaHSO_3$, 10% (v/v) glycerol
- Ammonium sulfate (AS)
- Buffer A: 50 mM Tris-HCl, pH 8.0, 200 mM NaCl, 10 mM 2-mercaptoethanol, 10% (v/v) glycerol
- Buffer B: 5 mM Tris-HCl, pH 8.0, 20 mM NaCl, 1 mM 2-mercaptoethanol, 1% (v/v) glycerol
- Buffer C: 50 mM Tris-HCl, pH 8.0, 1 M NaCl, 10 mM 2-mercaptoethanol, 10% (v/v) glycerol
- DNA-cellulose prepared according to Alberts and Herricke (8) and equilibrated in buffer A
- Gauze
- Immersible CX-10 filtration unit (Millipore) or equivalent
- Polymerase buffer: 200 mM Tris-HCl, pH 8.0, 200 μM NAD, 20 mM $MgCl_2$, 2 mM DTT, 2 mM PMSF, 20% (v/v) ethanol, 20 μg/ml histone 1-4, 20 μg/ml calf thymus DNA
- Deoxyribonuclease I (EC 3.1.4.5) stock solution: 1 mg/ml in 50 mM Tris-HCl, pH 8.0
- Perchloric acid (PCA)
- 1 M NaOH/30 mM NaEDTA
- 96% ethanol and 70% ethanol
- Affi-Gel 601 (Bio-Rad)
- 0.5 M Tris-HCl, pH 6.8
- 0.2 M ammonium bicarbonate
- Regulated water bath

Protocol 8.1

Purification of poly(ADP-ribose) polymerase

1. All subsequent procedures are performed at 4 °C. Homogenize 200 g of fresh calf thymus in 600 ml of extraction buffer using a blender (2 × 15 sec). Filtrate the homogenate through 1 layer of gauze and centrifuge at 9000 × g for 20 min. Discard the pellet.
2. Add solid ammonium sulfate (AS) to 30% saturation, stir for 30 min, and centrifuge at 9000 × g for 30 min. Discard the pellet. Adjust to 70% AS saturation, stir for 90 min, and centrifuge at 9000 × g for 45 min. Discard the supernatant.
3. Dissolve the precipitate in 80 ml of buffer A, dialyze against buffer B until the conductivity of the protein solution corresponds to that of buffer A. Add 100 ml of DNA-cellulose and incubate for 90 min (shaking slightly). Centrifuge at 300 x g for 4 min, discard the supernatant, and wash the DNA-cellulose twice with buffer A.
4. Elute the enzyme by adding 30 ml of buffer C to DNA-cellulose and adjusting the conductivity of the suspension with 5 M NaCl to that of buffer C. Shake slightly for 15 min and centrifuge at 300 x g for 4 min. Repeat the elution procedure with another 30 ml of buffer C and pool the supernatants. Store the enzyme fraction in liquid nitrogen and use it within 6 months.

Protocol 8.2

Preparation of poly(ADP-ribose)

1. Concentrate the above 60 ml of poly(ADP-ribose) polymerase fraction to a volume of 12 ml, using Immersible CX-10 filtration units. Add 30 ml of polymerase buffer and adjust with distilled water to a final volume of 60 ml. Add 60 µl of DNase I stock solution, vortex vigorously, and incubate the reaction mixture in a water bath at 30 °C for 30 min.
2. Digest the DNA by adding 600 µl of DNase I stock solution and incubating at 30 °C for 15 min.
3. Precipitate the ADP-ribosylated proteins by adding ice-cold PCA (final concentration 10% v/v), keep at 0 °C for 30 min, centrifuge at 1000 × g for 10 min, wash the pellet twice with 70% ethanol, and dry in a desiccator.

4. For cleavage of the covalent poly(ADP-ribose)-protein linkage dissolve the pellet in 6 ml of 1 M NaOH containing 30 mM NaEDTA. Incubate at 60 °C for 60 min, chill the sample to 0 °C, add 660 µl of acetic acid, ethanol precipitate the ADP-ribose polymers at –20 °C overnight, centrifuge at 1000 × g for 30 min, and dry in a desiccator.
5. Dissolve the pellet in 20 ml of 0.2 M ammonium bicarbonate. Apply the sample to an Affi-Gel 601 column (30 ml), wash with 0.2 M ammonium bicarbonate, and elute the ADP-ribose polymers with 0.5 M Tris-HCl, pH 6.8, observing the extinction at 260 nm.
6. Ethanol precipitate the ADP-ribose polymers at –20 °C overnight, centrifuge at 15 000 × g for 30 min, rinse the pellet with 70% ethanol, and dry in a desiccator. Dissolve the pellet [approximately 700 µg of poly(ADP-ribose)] in 1 ml of distilled water, determine the poly(ADP-ribose) concentration (1 A_{260} = 35.26 µg/ml) and adjust to 0.5 µg/µl. Store at –20 °C.

Biotinylation of poly(ADP-ribose)

Materials and buffers

- *N*-(4-Azido-2-nitrophenyl)-*N*'-(3-biotinylaminopropyl)-*N*'-methyl-1,3-propanediamine (photobiotin) stock solution: 1 mg/ml in distilled water (Sigma)
- 50 mM Tris-HCl, pH 9.0
- 2-Butanol
- Gas discharge lamp (Sylvania HSL-BW 400 W, Osram MB/U 400 W or equivalent)
- Ice-water bath
- Siliconized centrifuge tubes or beaker

Protocol 8.3

Biotinylation of poly(ADP-ribose)

1. Mix 200 µl of poly(ADP-ribose) solution [100 µg of poly(ADP-ribose)] with 33 µl of photobiotin stock solution and adjust with water to a final volume of 330 µl. The level of the solution in an open vessel should be 2–3 mm. Cool the solution in an ice-water bath and irradiate with the gas-discharge lamp from the top at a distance of 10 cm for 20 min.
2. Add 50 mM Tris-HCl, pH 9.0, to a final volume of 2 ml. Add 2 ml of 2-butanol, shake vigorously (vortex), and separate the 2 phases in a microcentrifuge. Discard the upper phase (2-butanol) containing the surplus photobiotin that has not reacted with poly(ADP-ribose). Repeat the 2-butanol extraction. Adjust the reduced volume of the aqueous phase with water to 1 ml. Ethanol precipitate the biotinylated poly(ADP-ribose) at –80 °C for 2 hours and centrifuge at 20 000 × g for 30 min. Wash the pellet with 70% ethanol, air-dry, and dissolve it in 200 µl of distilled water.

Photobiotin is light sensitive! Before irradiation all procedures must be carried out in very subdued light (darkroom).

Use of biotinylated poly(ADP-ribose) for ligand blotting

Materials and buffers

- Samples of poly(ADP-ribose) binding protein(s)
- SDS-polyacrylamide gel electrophoresis (PAGE) unit for separating proteins
- Electroblotting unit for transferring separated proteins to nitrocellulose
- Blocking solution: 5% bovine serum albumin (BSA) in Buffer I
- Buffer I: 50 mM Tris-HCl, pH 8.0, 0.5 M NaCl
- Buffer II: 10 mM Tris-HCl, pH 8.0, 0.15 M NaCl, 10 mM $MgCl_2$, 1 mM DTT

- Streptavidin-alkaline phosphatase conjugate (Amersham) diluted 1:2000 in 100 mM Tris-HCl, pH 7.4, 0.5 M NaCl, 2 mM $MgCl_2$, 0.05% Triton-X 100
- Buffer III: 100 mM Tris-HCl, pH 9.5, 100 mM NaCl, 5 mM $MgCl_2$
- Nitro blue tetrazolium (NBT) stock solution: 50 mg/ml in 70% dimethylformamide
- Bromochloroindolyl phosphate (BCIP) stock solution: 50 mg/ml in dimethylformamide

Protocol 8.4

Biotinylated poly(ADP-ribose) for ligand blotting

1. Separate proteins by SDS-PAGE and transfer onto nitrocellulose membranes (for standard procedures see ref. 9).
2. Treat the membrane with blocking solution for 1 hour, wash twice with large amounts of buffer I, and incubate with 60 ng/ml of biotinylated poly(ADP-ribose) in buffer II (0.25 ml/cm^2 membrane) at 22 °C overnight.
3. Wash the membrane 3 times with an excess of buffer II and incubate with diluted streptavidin-alkaline phosphatase conjugate (0.25 ml/cm^2 membrane).
4. Wash the membrane 5 times with an excess of buffer III. Develop the blot by incubating with 0.3 mg/ml of NBT, 0.15 mg/ml BCIP in buffer III for about 10 min. Stop the color reaction by washing the membrane with distilled water.

8.3 Results and Discussion

Table 8.1 shows that the biotinylation of poly(ADP-ribose) reaches a plateau at a photobiotin:poly(ADP-ribose) ratio of approximately 0.3. Accordingly, incubation of poly(ADP-ribose) with photobiotin concentrations above that level is not recommended, as this would only slightly increase biotinylation while unnecessarily increasing the amount of unreacted substrate.

Table 8.1 Biotinylation of poly(ADP-ribose) at increasing photobiotin concentrations

Ratio of photobiotin: poly(ADP-ribose)	Integrated O.D. (x 10^2)
0.03	12
0.15	90
0.3	140
1.5	150

Ten-microgram aliquots of poly(ADP-ribose) were labeled at different photobiotin concentrations. Final volume was 40 µl. Other conditions correspond to those given in the text. Dot blots of biotinylated poly(ADP-ribose) were stained by the streptavidin-alkaline phosphatase assay. The densitometric signals were integrated by the BioImage system of Millipore.

Under the conditions presently given, amounts of poly(ADP-ribose) as low as 0.1 ng were detected by the streptavidin-alkaline phosphatase-mediated color reaction. Certainly, the use of novel chemiluminescent substrates such as 4-methoxy-4-(phosphatephenyl)spiro-(1,2-dioxetane)-3,2'-adamantane (Lumigen^{M}) should further increase the sensitivity.

Apparently, the protein-binding properties of poly(ADP-ribose) were not affected by its covalent attachment to biotin. Figure 8.1 shows that the histone bands detected with biotinylated poly(ADP-ribose) corresponded to those detected with unmodified radioactive poly(ADP-ribose). Similarly, the biotinylated probe enabled the detection of poly(ADP-ribose)-binding proteins in a complex mixture of nuclear proteins (Fig. 8.2).

Figure 8.2 also points to the necessity for control experiments if crude preparations of proteins are analyzed (see also Troubleshooting). As seen in lane 2, some protein might give a signal by interacting with the streptavidin-alkaline phosphatase complex.

Summarizing, use of commercially available photobiotin with its photoreactive aryl-azido group proved useful for the modification of poly(ADP-ribose). This rapid and efficient labeling procedure can be recommended for small- and large-scale preparation of biotinylated poly(ADP-ribose).

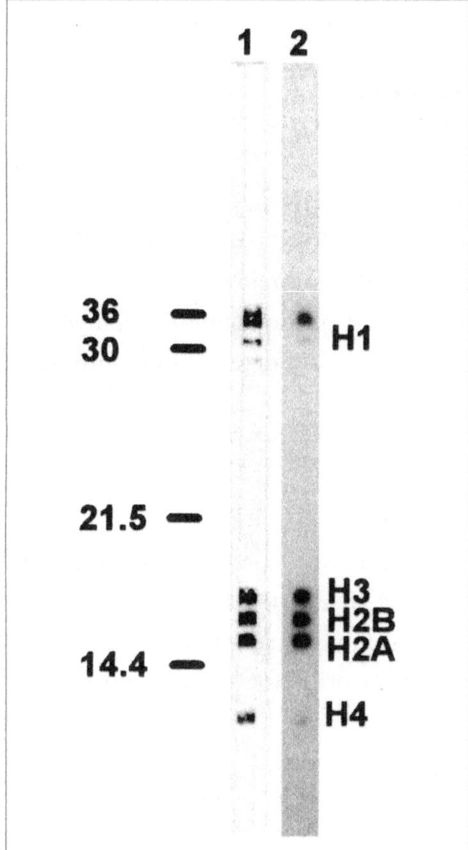

Figure 8.1 Use of biotinylated poly(ADP-ribose) for ligand blotting

Histones were electrophoretically separated by 15% SDS-PAGE and blotted onto nitrocellulose. Poly(ADP-ribose) was labeled with photobiotin as described in the text. Molecular masses are in kDa as indicated. Lane 1, incubation with biotinylated poly(ADP-ribose) and detection by streptavidin-alkaline phosphatase assay; lane 2, incubation with poly([P^{32}]ADP-ribose) and detection by autoradiography.

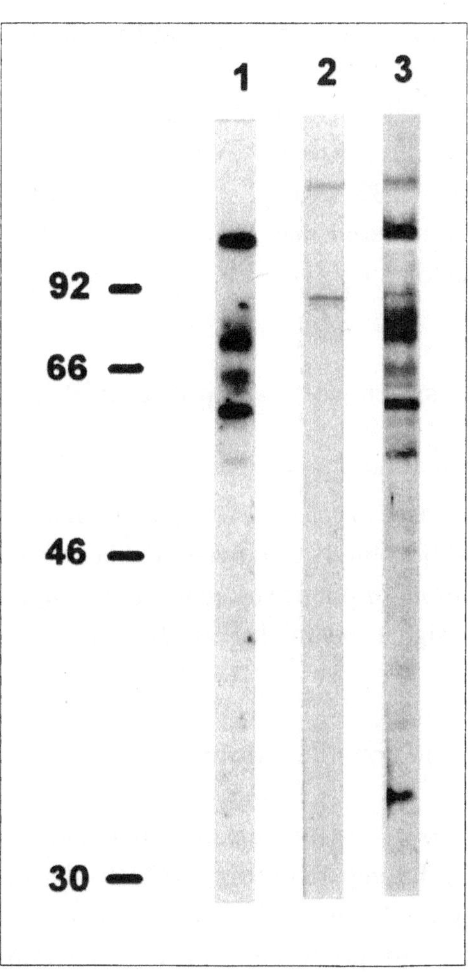

Figure 8.2 Detection of nuclear poly(ADP-ribose)-binding proteins

Proteins of a rat liver nuclear extract were separated by 12% SDS-PAGE and blotted onto nitrocellulose. Lane 1, incubation with poly([P32]ADP-ribose) and detection by autoradiography; lane 2, control, incubation with the streptavidin-alkaline phosphatase assay system solely; lane 3, incubation with biotinylated poly(ADP-ribose) and detection by the streptavidin-alkaline phosphatase assay.

8.4 Troubleshooting

Purification and storage of poly(ADP-ribose) polymerase

It is highly recommended that fresh calf thymus be used to prepare the enzyme. Adjustment of the conductivity of buffer C is essential for the quantitative elution of poly(ADP-ribose) polymerase from the DNA-cellulose matrix.

In our experience, the partially purified poly(ADP-ribose) polymerase lost approximately 50% of its original activity during 6 months' storage in liquid nitrogen.

Preparation of poly(ADP-ribose)

For additional purification of poly(ADP-ribose) a 2-hour incubation with 5 mg/ml proteinase K in 100 mM Tris-HCl, pH 7.4, 150 mM NaCl, 10 mM NaEDTA, 0.5% SDS at 37 °C may be included after the PCA precipitation step (4). Furthermore, the material may be extracted with phenol/$CHCl_3$ prior to Affi-Gel 601 chromatography (4).

Unspecific signals at ligand blotting

When partially purified protein fractions are analyzed, control experiments should be made to detect any unspecific signal. Thus, individual control blots not incubated with biotinylated poly(ADP-ribose) should undergo the detection procedure. Conditions of higher stringency (e.g., higher salt concentration) will prevent unspecific signals but might also affect the extent of ligand-protein binding.

Time considerations

Allow 1 day for the purification of the poly(ADP-ribose) polymerase and $2^1/_2$ days for the preparation of poly(ADP-ribose). Biotinylation of the poly(ADP-ribose) requires 4 hours. These labour-intensive steps need be done only every few months, since the product is stable if stored properly. The time for the protein separation by

SDS-PAGE and for the transfer to the membranes depends on the technical system used. Ligand blotting requires overnight incubation and an additional 3 hours to obtain the final results.

Acknowledgments

Grateful acknowledgment is made to Academic Press for permission to present data previously published in *Analytical Biochemistry* **220**, 415–419 (1994).

References

1 Ueda, K. and Hayaishi, O. (1985) *Ann. Rev. Biochem.* **54**, 73–100.

2 Jacobson, M.K. and Jacobson, E.L. (1989) *ADP-Ribose Transfer Reactions: Mechanisms and Biological Significance.* Springer, New York.

3 Boulikas, T. (1991) *Anticancer Res.* **11**, 489–528.

4 Wesierska-Gadek, J. and Sauermann, G. (1988) *Eur. J. Biochem.* **173**, 675–679.

5 Nozaki, T., Masutani, M., Akagawa, T., Sugimura, T. and Esumi, H. (1994) *Biochem. Biophys. Res. Commun.* **198**, 45–51.

6 Narendja, F.M. and Sauermann, G. (1994) *Anal. Biochem.* **220**, 415–419.

7 Forster, A.C., McInnes, J.L., Skingle, D.C. and Symos, R.H. (1985) *Nucl. Acids Res.* **13**, 745–761.

8 Alberts, B. and Herrick, H. (1971) *Methods Enzymol.* **21**, 198–217.

9 Sambrook, J., Fritsch, E.F. and Maniatis, T. (1989) *Molecular Cloning.* Cold Spring Harbor Laboratory Press, New York.

Preparation and Use of Biotinylated Probes for the Detection and Characterisation of Serine Proteinase and Serine Proteinase Inhibitory Proteins

James Melrose and
Kenneth J. Rodgers

Summary

Trypsin (1), secretory leucocyte proteinase inhibitor (SLPI) (2), potato chymotrypsin inhibitor-1 (PCTI-1) (3), and aprotinin (4) were successfully biotinylated with retention of their biological activities. Biotin-labeled SLPI was suitable as a standard in an ELISIA for the quantitation of human connective tissue serine proteinase inhibitors (SPIs) (2). Biotin-labeled PCTI-1 and aprotinin were sensitive probes for the detection of a range of serine proteinases on Western blots including chymotrypsin, plasmin, leucocyte elastase, cathepsin G, and trypsin and were also used to demonstrate the synthesis of a novel "chymotrypsin-like" proteinase by ovine chondrocytes in culture (3, 4). Biotinylated trypsin (bT) was a sensitive probe on Western blots for the detection and characterisation of a range of serum and connective tissue-derived SPIs. The high sensitivity afforded by the bT detection system facilitated the characterisation of an endogenous SPI extracted from ovine articular cartilage and synthesised by ovine chondrocytes in culture (5). It should be noted that the biotinylated probes only detect biologically active functional proteins, a particularly appropriate parameter to monitor in disease states. Furthermore, conventional Western blotting employing specific antibodies or specific inactivation of selected proteinases using chloromethylketones may be undertaken on duplicate blots to complement the biotinylated probe data.

9.1 Introduction

The affinity of avidin for biotin is one of the strongest noncovalent interactions known in nature ($K_a = 10^{-15}$ M) (6–9). Furthermore, this interaction is rapid, and once formed, the avidin-biotin complex is relatively unaffected by extremes in pH, temperature, organic solvents and other denaturing agents (6–9). Biotin is a small molecule (244 Da) and can be conjugated to macromolecules using a range of chemistries without significantly altering the size, physical characteristics, or biological activity of the macromolecule in question (6–9). Biotin labeling therefore represents an extremely useful means of monitoring specific biomolecules in complex biological systems, since the biotinylated molecule may be specifically and sensitively detected using avidin conjugated to a reporter group. This versatile approach has been used in a diverse range of bioanalytical applications (6–10), and in this chapter we describe how we have applied the biotin-avidin interaction for the detection of a range of serine proteinases and their inhibitory proteins.

In the present study, 2 basic strategies were adopted for the biotinylation of serine proteinases and SPIs, (i) direct biotin labeling of the native protein, or (ii) labeling of an enzyme-inhibitor complex in order to prevent the active site(s) of the enzyme or SPI from becoming labeled, since this could potentially sterically impede the subsequent interaction of the enzyme or SPI with its complementary partner. An important consideration was that the biotin-labeled probes prepared in this study should be functionally similar to their nonlabeled counterparts. Biotinylated probes were therefore prepared using a range of incorporation levels of biotin, and relevant compositional and functional analyses of the nonlabeled and biotinylated probes were undertaken side by side to establish the most suitable probe for a particular application. For example, SLPI and bSLPI had similar association constants for solid phase trypsin and chymotrypsin, and the bSLPI could therefore be used as a standard in a competitive-binding, enzyme-linked immunosorption inhibition assay (ELISIA) for the quantitation of SLPI-like SPIs in human connective tissues (2). The other main area of application of the biotinylated probes in the present study was their use in the detection of SPIs or serine proteinases on Western blots following gradient SDS-polyacrylamide gel electrophoresis (PAGE). Biotin-labeled synthetic proteinase inhibitory compounds such as biotinyl-aminoacyl-chloromethanes (11, 12), biotinyl-isocoumarins (13) or biotinylated peptides (14) have also been synthesised (15) and shown to have application in the identification and characterisation of a range of serine and thiol-proteases. The biotinylated probes used in the present study are prepared from common biochemicals by well-established methodologies using equipment which should be readily available in most biochemistry laboratories. Moreover, this study demonstrated that the sensitivity and specificity attainable using such biotinylated protein probes was comparable to the aforementioned biotinyl chloromethanes and isocoumarin derivatives (11, 12, 15).

9.2 Technical Procedures

Materials

The sources of key materials only are given; details of other biochemicals used in the applications are given elsewhere (1–4). Bovine pancreatic trypsin (Type XIII, TPCK treated), SBTI (Sigma, type 1-S), 4-nitrophenyl 4'-guanidinobenzoate (NPGB), 2-hydroxy-5-nitro-α-toluene-sulphonic acid sultone (HNTSS), CBZ-arginine-4-nitroanilide (ZAPNA), succinyl-Ala-Ala-Pro-Phe-4-nitroanilide (SAAPPNA), D-biotin, avidin (14.8 units/mg protein), immobilised avidin (monomeric) on 4% beaded agarose, 2-(4'-hydroxyazobenzene)-benzoic acid (HABA), biotin succinimide ester (NHS-d-biotin), avidin-labeled alkaline phosphatase conjugate (0.5 mg protein/ml, 3.8 units avidin/mg protein, 390 units alkaline phosphatase/mg protein) were purchased from Sigma (St. Louis, MO, USA). Trasylol (10 000 Kallikrein inhibitory units/ml; ~2×10^{-4} mol/l; ~1.38 mg/ml) was purchased from Bayer Australia (Pymble, NSW, Australia). Sulfosuccinimidyl-6-(biotinamido) hexanoate (NHS-LC-biotin) was purchased from Pierce (Rockford, IL, USA). The SLPI used in this study was a kind gift of Dr R.C. Thompson, Synergen, Boulder, CO, USA (28), this protein is available commercially from R & D Systems (Minneapolis, MN, USA). Precast 4–20% polyacrylamide (Tris-glycine), gradient slab gels (0.1 × 10 × 10 cm) nitrocellulose blotting membranes, prestained and broad-range electrophorcsis standard proteins were purchased from Novex (Frenchs Forest, NSW, Australia).

Biotinylation procedures

Procedure 9.1

Biotinylation of SLPI

Aliquots of biotin succinimide ester (NHS-D-biotin, 17 mM) in dimethyl sulfoxide (DMSO) were mixed in various molar ratios (1:1, 2:1, 5:1, 10:1) with SLPI (0.17 mM) in 100 mM sodium acetate, pH 5.0 (1.1 ml) (2). After mixing at room temperature for 60 min, 50 μl of 1 M NH_4Cl was added to quench unreacted succinimide ester, and after a further 10 min the samples were dialysed exhaustively against phosphate-buffered saline (PBS, pH 7.2) (4 × 4 l × 12 hours) to remove free biotin.

Procedure 9.2

Biotinylation of PCTI-1

A 20-fold molar excess of NHS-LC-biotin was reacted with a solution of PCTI-1 (0.1 mg/ml) in PBS, pH 7.2, for 3 hours at room temperature. An aliquot of 1 M NH_4Cl was then added (0.5 ml), and after 10 min the sample was exhaustively diafiltrated over a YM2 (2 kDa cutoff) membrane to exchange the buffer for Tris-buffered saline (TBS) and to remove any free NHS-LC-biotin.

Procedure 9.3

Biotinylation of aprotinin

A 10-fold molar excess of NHS-LC-biotin was mixed with a solution of aprotinin (trasylol, 1 mg/ml) in PBS (pH 7.2, 20 ml) at room temperature for 5 hours, 0.5 ml of 1 M NH_4Cl was then added, and the sample was diafiltrated as indicated above to remove excess NHS-LC-biotin.

Procedure 9.4

Biotinylation of aprotinin-trypsin complex

Aprotinin (trasylol, 28 mg, 4.7 μmol) and bovine pancreatic trypsin (100 mg, 4.3 μmol) in PBS (pH 7.8, 30 ml) were mixed end-over-end for 30 min at room temperature. Biotin succinimide ester (NHS-D-biotin) in DMSO (29.4 mg, 86 μmol/3 ml) was then added, and biotinylation was allowed to proceed for 6 hours at room temperature. An excess of NH_4Cl was added, and the sample was diafiltrated as indicated above to remove free biotin.

Procedure 9.5

Biotinylation of trypsin

Biotin succinimide ester (NHS-d-biotin, 17 mM) in DMSO was mixed in various molar ratios (2:1, 5:1, 10:1) with trypsin (0.213 μmol) in PBS (pH 7.2, 2.0 ml) containing DMSO (12.5% v/v) for 3 hours at room temperature; 0.1 ml of 1 M NH_4Cl was then added, and the sample was dialysed against 0.1 M glycine-HCl, pH 2.0, containing 1 M NaCl to remove free nonreacted biotin.

Procedure 9.6

Biotinylation of trypsin-SBTI complex

Soybean trypsin inhibitor (SBTI, 1.5 μmol) and trypsin (0.5 μmol) in PBS (pH 7.2, 2.0 ml) were mixed end-over-end for 30 min at room temperature to form trypsin-SBTI complex. Biotin succinimide ester dissolved in DMSO (0.2 ml/20 μmol) was added, and the sample was mixed at room temperature for an additional 3 hours. An aliquot of 1 M NH_4Cl (0.5 ml) was added, and after an additional 10 min the sample was dialysed against 0.1 M glycine-HCl, pH 2.0, containing 1 M NaCl.

Procedure 9.7

Isolation of biotin-labeled proteins

Avidin affinity chromatography

Biotinylated samples equilibrated in TBS and free of nonreacted biotin succinimide ester were purified by avidin affinity chromatography (16, 17). The avidin-monomer affinity column (10-ml bed volume) was initially treated with 2 mM biotin in TBS (pH 7.2, 50 ml) to block any high-affinity binding sites. Low-affinity biotin-binding sites were then exposed by elution with 0.1 M glycine-HCl buffer (pH 2.0, 50 ml), and the column was reequilibrated in TBS, pH 7.2 (50 ml). Aliquots of biotinylated proteins (~1–2 mg) were applied in TBS pH 7.2, and nonbound material was eluted with TBS (pH 7.2, 50 ml), prior to elution of bound material with 2 mM biotin in TBS, pH 7.2 (50 ml). Fractions of 4.0 ml were collected at a flow rate of 20 ml/hour and monitored for protein using the bicinchoninic acid procedure (18) and inhibitory activity against bovine pancreatic trypsin using ZAPNA or against bovine pancreatic chymotrypsin using SAAPPNA as substrates (19, 20).

Protocol 9.8

Separation of free biotin from biotinylated proteins by gel-permeation chromatography

Free biotin was separated from avidin affinity-purified biotinylated aprotinin (bA), PCTI-1, and SLPI samples, equilibrated in TBS using Sephadex-G10 chromatography. These samples were applied to a column of Sephadex-G10 (1.6 × 84 cm^2) eluted with TBS at 120 ml/hour. Fractions of 2 ml were collected, and aliquots were monitored for protein (18), trypsin- or chymotrypsin-inhibitory activity (as appropriate) (19, 20), and for biotin by displacement of avidin from preformed avidin-HABA complex (21). Samples of avidin-purified biotinylated aprotinin-trypsin complex (215 nmol of aprotinin and trypsin) or SBTI-trypsin (0.2 μmol of trypsin, 0.6 μmol of SBTI) equilibrated in 0.1 M glycine-HCl buffer, pH 2.0, containing 1 M NaCl were applied to columns of Sephadex-G25 (1.6 × 96 cm^2) or Sephacryl S100HR (1.6 × 85 cm^2), respectively, equilibrated in the same buffer. Fractions (0.5 ml) were collected at a flow rate of 30 ml/hour or 60 ml/hour, respectively. Fractions were monitored for protein, trypsin-inhibitory and trypsin activity, and for biotin as indicated above. Aliquots of selected Sephadex-G25 or Sephacryl S100HR fractions were examined by 4–20% or 10–20% polyacrylamide gradient SDS-PAGE (22, 23) to confirm the separation of trypsin, aprotinin, and SBTI from one another. Pooled bA fractions deemed to be free of trypsin were made 1 mM in AEBSF to inactivate any residual trypsin activity, concentrated by diafiltration over a YM2 membrane, and stored at 4 °C. Pooled bT fractions free of SBTI were purified further by SBTI affinity chromatography.

Procedure 9.9 **SBTI affinity chromatography of biotinylated trypsin**

Biotinylated trypsin samples (0.064 μmol) from avidin affinity and Sephacryl S100HR gel-permeation chromatography were equilibrated in 50 mM Tris-HCl buffer, pH 7.5, containing 0.15 M NaCl by diafiltration and applied at 4 °C to a column (1-ml bed volume) of SBTI-Sepharose 4B. The column was washed at 20 ml/hour (1-ml fractions) with 50 mM Tris-HCl buffer, pH 7.5, containing 0.15 M NaCl (10 ml) to remove nonbound material. One milliliter of 5 mM HCl, pH 2.2, containing 1 M NaCl and 10 mM $CaCl_2$ was then run into the column, and the flow was stopped for 10 min. Elution was then resumed with the same buffer (19 ml) to elute the biotinylated trypsin. Aliquots of fractions were monitored for protein and trypsin activity against ZAPNA (19), and also examined by SDS-PAGE (22).

Compositional and functional analyses conducted on the biotinylated probes

The absolute biotin content of purified biotinylated samples and the relative biotin content of chromatography fractions were determined colorimetrically by displacement of avidin from avidin-HABA complex (21). For the measurement of the relative biotin content of chromatographic fractions, aliquots of eluant fractions (0.1 ml) were mixed with avidin-HABA complex (0.2 mg/ml and 0.25 mM, respectively) in PBS (pH 7.2, 0.2 ml) in 96-well micro titreplates, and the reduction in absorbance at 490 nm relative to control samples was measured by plate reader. The absolute biotin content of purified biotinylated samples was determined spectrophotometrically (21) using the λ_{max} avidin-HABA complex as 500 nm, and the molar extinction coefficient $\varepsilon_{500\,nm} = 35\,500 \times M^{-1} \times cm^{-1}$. In some samples the number of free amino groups modified by biotinylation of the probe was also determined (24). The specific activity and IC50 values of nonmodified and biotinylated SLPI, PCTI, and aprotinin samples were determined by titration against active site-titrated trypsin or chymotrypsin (25), respectively. The protein contents of the SLPI, PCTI-1, and aprotinin samples were measured by the bicinchoninic acid procedure (18) using nonmodified SLPI, PCTI-1, or aprotinin as standards. The concentration of standard aprotinin samples was verified spectrophotometrically (26) as-

Probe	Application	Comments	Reference
bT	Detection of α_1-PI, inter-α-trypsin inhibitor, SBTI variants, aprotinin, and SLPI from human connective tissues on Western blots with high sensitivity.		Melrose, J., Rodgers, K. and Ghosh, P.(1992) *Anal. Biochem.* **222**, 34–43.
	Identification and characterisation of a novel serine proteinase inhibitor synthesised by ovine chondrocytes.	Detection sensitivity α_1PI 25 fmol trasylol 23 fmol	Rodgers, K., Melrose, J. and Ghosh, P. (1996) *Electrophoresis* ***in press***. **17**, 213–218
	Identification and characterisation of serine proteinase inhibitory proteins on nitrocellulose squash blots of isoelectric focussing gels		
bSLPI	Enzyme-linked immunosorbent inhibition assay (ELISIA) for the detection of SLPI-like serine proteinase inhibitory proteins from extracts of human intervertebral discs.	2 ng of SLPI detected	Melrose, J. and Ghosh, P. (1992) *Anal. Biochem.* **204**, 372–382.
bA	Detection of plasmin, leucocyte elastase, cathepsin G, trypsin, chymotrypsin on Western blots with high sensitivity and specificity. Identification and characterisation of an ovine chondrocyte chymotrypsin-like serine proteinase.	0.2 ng of trypsin, chymotrypsin detected	Melrose, J., Ghosh, P., and Patel, M. (1995) *Int. J. Biochem. Cell Biol.* ***in press***. **27**, 891–904.
bPCTI-1	Identification and quantitation of chymotrypsin-like serine proteinases on Western blots. Identification and characterisation of a chymotrypsin-like serine proteinase synthesised by ovine chondrocytes in culture.	0.1 ng of chymotrypsin detected	Rodgers, K., Melrose, J. and Ghosh, P. (1995) *Anal. Biochem.* **227**, 129–134.

suming that a 1-mg/ml solution had an A280 nm (1 cm) value of 0.84. Nonmodified and biotinylated trypsin samples were active site-titrated using NPGB (25), their protein contents were measured by the BCA procedure (18) using bovine pancreatic trypsin as standard, and their specific activities were calculated.

Procedure 9.10

Determination of trypsin and trypsin inhibitory activities

Trypsin was assayed at 37 °C with the synthetic substrate ZAPNA (19) in 50 mM Tris-HCl buffer, pH 8.2, containing 0.1 M NaCl, 10mM $CaCl_2$, 0.1 mg/ml BSA, and 0.02% NaN_3. Trypsin-inhibitory activity was calculated from the decrease in absorbancy of samples at 405 nm compared with trypsin control samples measured using an automatic plate reader.

Procedure 9.11

Determination of chymotrypsin inhibitory activity:

Chymotrypsin inhibitory activity was measured against bovine pancreatic chymotrypsin using SAAPPNA (20) as substrate at 25 °C in 50 mM Tris-HCl buffer (pH 7.8) containing 0.1 M NaCl, 10 mM $CaCl_2$, 0.1 mg/ml BSA, and 0.02% w/v NaN_3. Inhibitory activity was calculated from the decrease in absorbancy at 405 nm compared with chymotrypsin control samples measured using an automatic plate reader.

Calculation of IC50 values

The amount of bA and bPCTI-1 providing 50% inhibition (IC50) of a standard amount of active-site-titrated trypsin or chymotrypsin was measured by titration of variable amounts of bA and bPCTI-1 with standard amounts of trypsin or chymotrypsin and determination of the residual trypsin or chymotrypsin activity as indicated above. A semi-log plot of residual enzyme activity versus percentage of inhibitory activity was constructed, and IC50 values were determined from this data (2, 4).

◀ ***Table 9.1 Applications of biotinylated trypsin (bT), and biotinylated SLPI (bSLPI), aprotinin (bA), and potato chymotrypsin inhibitor-1 (bPCTI-1).***

Procedure 9.12

Active-site titration of trypsin or chymotrypsin

Trypsin: Active-site titration of trypsin is conveniently undertaken using 4-nitrophenyl-4-guanidinobenzoate (4-NPGB) (25). Trypsin solutions (0.7–3.5 mg/ml) in 1 mM HCl containing 20 mM $CaCl_2$ are stored aliquoted at –20 °C; 4-NPGB (hydrochloride salt) (16.8 mg) is dissolved in 1 ml of dimethylformamide, then diluted into 4 ml of acetonitrile and stored at 4 °C for up to 1 week. The most suitable assay buffer for active site titration is 0.1 M veronal buffer (pH 8.3) containing 20 mM $CaCl_2$ (2.06 g of sodium diethylbarbiturate and 294 mg of $CaCl_2$ in 90 ml of water adjusted to pH 8.3 with dilute HCl, then the total volume adjusted to 100 ml). To a 1-cm-path-length, 1-ml spectrophotometer cuvette are added veronal buffer (0.8 ml) and trypsin (0.2 ml), and a baseline A410 nm reading is taken. 4-NPGB (5 μl) is added with efficient mixing to start the reaction, and a second A410 nm reading is taken after 10 sec (π, ΔA410 nm). The concentration of active enzyme is calculated from the formula: $c = 3.027 \times 10^{-4} \times \pi$ mol/l (millimolar absorption coefficient for 4-nitrophenol $\varepsilon_{410} = 1.660 \text{ l} \times \text{mM}^{-1} \times \text{mm}^{-1}$ at pH 8.3).

Chymotrypsin: Active-site titration of chymotrypsin is undertaken using 2-hydroxy-5-nitro-α-toluene-sulphonic acid sultone (HNTSS) (25). Dissolve α-chymotrypsin (300 mg) in 0.1 M sodium acetate buffer, pH 5.0 (10 ml); HNTSS (64.56 mg) is dissolved in acetonitrile (100 ml). The assay buffer for titration is 0.1 M Tris buffer, pH 7.41 (1.21 g of Tris base dissolved in 80 ml of water is titrated to pH 7.41 with dilute nitric acid and adjusted to 100 ml). The following solutions are added to a 1-cm-path-length, 4-ml spectrophotometer cuvette: 0.1 M Tris buffer, pH 7.41 (3.0 ml), chymotrypsin (0.1 ml). The solutions are mixed thoroughly, and the A390 nm baseline is recorded. HNTSS (50 μl) is added to start the reaction, and after mixing again the A390 nm is recorded for 5 min. The ΔA390 nm (π) is used to calculate the active enzyme concentration from the formula $c = 3.788 \times 10^{-3} \times p$ mol/l (millimolar absorption coefficient of the sulphonyl enzyme is $\varepsilon_{390} = 0.8315 \text{ l} \times \text{mM}^{-1} \times \text{mm}^{-1}$ at pH 7.41).

Procedure 9.13

Modification of the active-site Ser 185 residue of bT using PMSF

Biotinylated trypsin (1 mg/ml) was made 4 mM in PMSF from a stock solution of 100 mM PMSF in absolute ethanol, and an additional aliquot of PMSF was added every 5 min for 20 min until the relative trypsin activity was <0.1% of the starting activity against the substrate *N*-benzoyl-Phe-Val-Arg-4-nitroanilide (19).The PMS-bT probe can be used in a manner analogous to bT on Western blots (1) and provides a useful way of distinguishing between the binding of active SPIs with bT (which bind to the bT active site and thus are not reactive with PMS-bT) and other proteins capable of binding to bT at sites removed from its active site (these are still reactive with PMS-bT).

Applications of the biotinylated probes (see also Table 9.1)

- Development of an avidin-biotin competitive inhibition assay and valida tion of its use for the quantitation of human intervertebral disc serine pro teinase inhibitory proteins (2)
- Preparation and use of biotinylated trypsin in Western blotting for the detection of trypsin inhibitory proteins (1)
- Biotin-labeled potato chymotrypsin inhibitor-1: a useful probe for the detection and quantitation of chymotrypsin-like serine proteinases on Western blots and for the detection of a serine proteinase synthesised by articular chondrocytes (3)
- Biotinylated aprotinin: a versatile probe for the detection of serine proteinases on Western blots (4)
- Biotinylated trypsin: a sensitive, versatile probe for the detection and characterisation of an ovine chondrocyte serine proteinase inhibitor using Western blotting (5)
- Identification of serine proteinase inhibitory proteins on squash blots of isoelectric focussing gels using a biotinylated trypsin probe for their detection

Table 9.2 Composition of biotinylated trypsin (bT), SLPI (bSLPI), aprotinin (bA), and PCTI-1 (bPCTI-1).

Probe	Mol biotin/mol probe	Number of free NH_2 groups modified(24)	%Probe labeled[§]	Specific activity (U/mg protein)	IC50 trypsin	IC50 chymotrypsin
bT from bT-A complex	5.0	4	96.8	60.2×10^{3}**	NA	NA
bT 2:1*	1.5	1	45.0	60×10^{3}**		
bT5:1*	4.0	3	85.0	60.1×10^{3}**	NA	NA
bT10:1*	6.0	4	94.5	58.2×10^{3}**		
bA from bA-T complex	1.2–1.5	-	92	4.6×10^{6}¶	3.2¶¶	
bA	1.8–2.0	–	94	4.4×10^{6}¶	2.5¶¶	–
bPCTI-1	2.9	–	90	3.9×10^{3}†	–	0.49††
bSLPI	2.0	2	95	–	2.29††	2.60††

§ proportion of probe which was bound to immobilised avidin, as protein

** molar ratio of succinimide ester used (1)*

*** One unit represents an increase in absorbance of 0.001 per minute using CBZ-Arg-4-nitroanilide as substrate in 100 mM Tris-HCl buffer (pH 8.2), protein measured by the BCA procedure (1, 18)*

¶ Kallikrein inhibitory units/mg protein (26)

¶¶ The amount of bA (nmol) required to provide 50% inhibition of active-site-titrated trypsin (5 nmol) (4)

† One inhibitory unit is the amount of bPCTI-1 (mg) required to reduce the rate of cleavage of SAAPP-NA by chymotrypsin (1 mg) to 50% (3)

†† The amount of bSLPI or bPCTI-1 (pmol) giving 50% inhibition of active-site-titrated trypsin or chymotrypsin (4.2 pmol) (2, 3)

NA, not applicable; –, analysis not undertaken

9.3 Results and Discussion

Biotinylated SLPI

Secretory leucocyte proteinase inhibitor (SLPI) is a 12-kDa, 107-amino acid serine proteinase inhibitor consisting of two domains (27) (Fig. 9.1). The elastase-, chymotrypsin- and trypsin-binding sites are located within the carboxyl terminal domain; 8 disulfide bridges maintain SLPI in an active conformation (27). Secretory leucocyte proteinase inhibitor contains 15 lysine residues, and thus is readily labeled with biotin succinimide esters. Biotinylation of SLPI is best undertaken at pH 5.0; the ε-amino groups of lysine (pK_a 8.95) are protonated but still reactive at pH 5, and SLPI is also more stable at pH 5.0 than the alkaline conditions more commonly used for biotinylation. Approximately 95% of the SLPI was labeled under the conditions employed as assessed by the proportion of the SLPI which could bind to an avidin affinity column (Table 9.2). Biotinylated SLPI containing 2 mol biotin/mol SLPI was found to have an IC50 similar to its nonlabeled counterpart and was subsequently found to be a sensitive standard in an ELISIA used for the quantitation of connective tissue SPIs; it could detect as little as 2 ng SLPI/well (2).

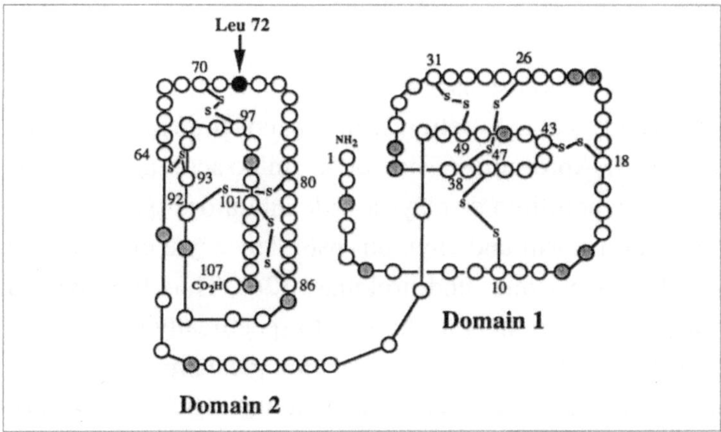

Figure 9.1 Simplified schematic of the primary structure and organisation
Demonstration of the distribution of lysine residues (stippled), which are primary targets during labeling of SLPI with biotin succinimide esters, and of the 8 disulfide bridges within the 2 domains of SLPI, which maintain it in an active conformation. The elastase-, trypsin-, chymotrypsin binding site is located in the carboxyl domain of SLPI (Leu 72, arrow).

Biotinylated PCTI-1

Chymotrypsin inhibitor-1 from potatoes (PCTI-1 or PPI-1) was first described by Melville and Ryan (28) and differs from the low molecular weight polypeptide potato chymotrypsin inhibitor (PCI-1) described by Haas et al. (29). PCTI-1 is a tetrameric protein of 38 kDa and is a relatively stable protein over the pH range 3–10 (28). Each of the subunits of PCTI-1 of $M_r \sim 9.5$ kDa contains a single reactive site for chymotrypsin and can be dissociated from the native SPI using GuH-Cl (2 M) or urea (8 M), but reassociates upon reintroduction into physiological buffers without loss in biological activity. In the present study the tetrameric form of PCTI-1 was biotinylated to minimise labeling of the chymotrypsin reactive sites of the monomeric SPI forms. The bPCTI-1 prepared in this study contained 3 mol of biotin per mol of 38 kDa PCTI-1 (see Table 9.2). Moreover, the IC50 of the labeled PCTI-1 was identical to its nonlabeled counterpart, demonstrating that no loss of inhibitory activity had occurred in the biotinylation procedure employed. Furthermore, bPCTI-1 was a very sensitive detection probe for chymotrypsin and chymotrypsin-like serine proteinases on Western blots (Fig. 9.2).

Biotinylated aprotinin

Aprotinin is particularly amenable to use as a detection probe, since it is a small protein (6512 Da), it is extremely stable to extremes in pH and temperature, and its primary structure is known (26). Aprotinin consists of 58 amino acids and contains 4 lysine residues. Furthermore, aprotinin displays a wide inhibitory spectrum and is known to inhibit pancreatic trypsin and chymotrypsin, leucocyte elastase and cathepsin G, kallikrein, plasma plasmin, and urokinase (26). Thus biotinylated aprotinin could potentially be an extremely useful tool to specifically monitor the aforementioned serine proteinases in a diverse range of biological processes. Since one of the lysine residues (lysine 15) is located adjacent to the active site of aprotinin, we prepared biotinylated aprotinin with the lysine-15 residue protected in an enzyme-inhibitor complex (aprotinin-trypsin) as well as the free aprotinin form.

Several peptidyl chloromethyl ketone derivatives with specificity for particular proteinases are available; these may be used to selectively inactivate proteinases on Western blots and thus aid in their subsequent identification using the bA detection

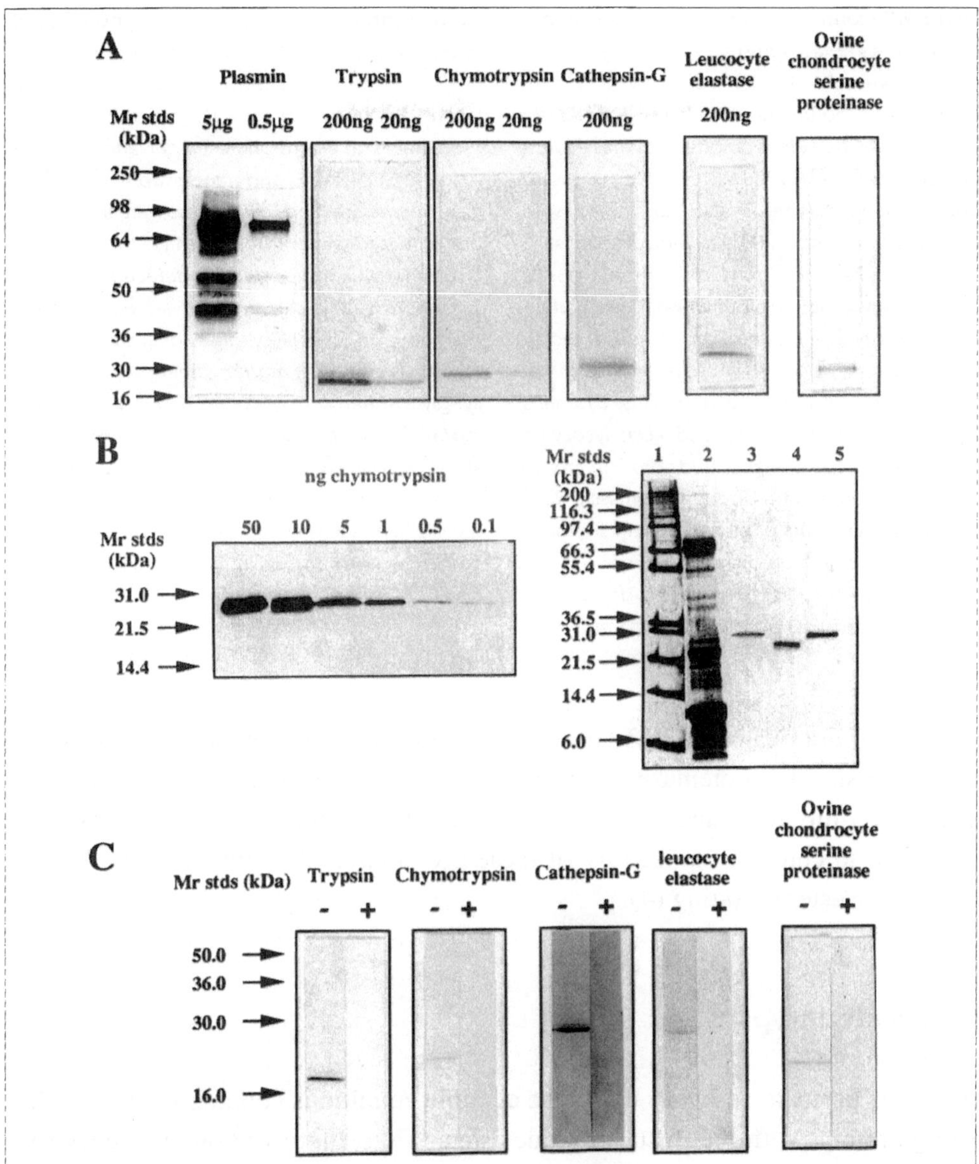

Figure 9.2 Detection of a range of serine proteinases

Detection on Western blots following 4–20% gradient (A, C) or 10–20% gradient SDS-PAGE (B) using (A) bA; (B) bPCTI-1 for detection; and (C) confirmation of the identities of the serine proteinases detected in (A) and (B) by pretreatment with specific chloromethylketones prior to the biotinylated probe detection step. The Novex prestained standards used in (A) and (C) were myosin (250 kDa), BSA (98 kDa), glutamic dehydrogenase (64 kDa), carbonic anhydrase (36 kDa), myoglobin (30 kDa), and lysozyme (16 kDa); these are indicated at the left-hand side of the figure. In (B), a range of chymotrypsin samples (0.1–5 ng/lane left-hand side of

Figure 9.2 (continued) *figure) or ovine chondrocyte media samples (right-hand side of figure) were electrophoresed on 10–20% polyacrylamide gradient gels using the Tris-tricine buffer system (23); the samples were then electroblotted to nitrocellulose and detected with bPCTI-1. The Novex broad-range protein standards also run in (B) were: myosin, rabbit muscle, 200 kDa; β-galactosidase, E. coli, 116.3 kDa; phosphorylase b, rabbit muscle, 97.4 kDa; bovine serum albumin, 66.3 kDa; glutamic dehydrogenase, bovine liver, 55.4 kDa; lactate dehydrogenase, porcine muscle, 36.5 kDa; carbonic anhydrase, bovine erythrocytes, 31.0 kDa; trypsin inhibitor, soybean, 21.5 kDa; lysozyme, chicken egg white, 14.4 kDa; and aprotinin, bovine lung, 6 kDa. Lanes 1 and 2 depict Coomassie R250-stained gel segments of Novex protein standards (as above) and a crude ovine chondrocyte media sample (25 μg of protein). Lanes 3–5 depict nitrocellulose electroblots of ovine chondrocyte media (25 μg of protein), bovine pancreatic trypsin (7.5 ng) or bovine pancreatic chymotrypsin (1 ng), respectively, which were detected using bPCTI-1 as probe (3). In (C), approximately 200 ng of the indicated proteinase was electrophoresed per lane by 4–20% gradient SDS-PAGE, and transferred to nitrocellulose. Respective lanes of the blots were then either preincubated without (–) or with (+) the following chloromethyl ketones prior to detection with bA. Trypsin: tosyl-lysyl-chloromethyl ketone, 1 mg/ml; chymotrypsin, CBZ-glycyl-glycyl-phenylalanyl-chloromethyl ketone, 0.1 mg/ml; cathepsin-G, CBZ-glycyl-glycyl-phenylalanyl-chloromethyl ketone, 0.1 mg/ml; human leucocyte elastase, succinyl-alanyl-alanyl-prolyl-alanyl-chloromethyl ketone, 0.1 mg/ml; ovine chondrocyte serine proteinase, CBZ-glycyl-glycyl-phenylalanyl-chloromethyl k etone, 0.1 mg/ml (modified from ref. 3 with permission from Anal. Biochem.).*

system (4). For example CBZ-Gly-Gly-Phe-chloromethyl ketone has affinity for chymotrypsin-like proteinases, and MeOSucc-Ala-Ala-Pro-Ala-chloromethyl ketone has affinity for leucocyte elastase. Both of these inhibitory compounds have been used to confirm the identity of these proteinases in combination with conventional Western blotting (4).

Biotinylated trypsin

Trypsin is prone to autolysis under the alkaline conditions commonly used for labeling proteins with biotin succinimide esters. Thus, biotinylation of free trypsin using NHS-d-biotin yielded very little active labeled trypsin (1). Conversely, biotinylation of trypsin with its active site protected in a reversible enzyme-inhibitor complex with either aprotinin or SBTI proved to be a superior means of labeling trypsin with biotin, and the bT could be recovered by gel-permeation, avidin, and SBTI affinity chromatographies in a highly active form (1). Biotinylated trypsin is also available commercially (Sigma, St. Louis, MO, USA) and is of a similar composition to the bT labeled in bT-aprotinin or SBTI complex in this study (1). Commercial bT should be further purified by SBTI affinity and stored aliquoted in 5 mM HCl (pH 3), 10 mM $CaCl_2$ at –20 °C. Biotinylated trypsin is an extremely sensitive probe for detecting SPIs on Western blots; and conventional Western blotting may also be undertaken using specific antibodies to further confirm the identity of the functionally active SPIs detected using bT (1, 3, 5). PMS-bT (Procedure 9.13) may also be used to confirm that the binding of bT to SPIs through their active sites (see Fig. 9.3C).

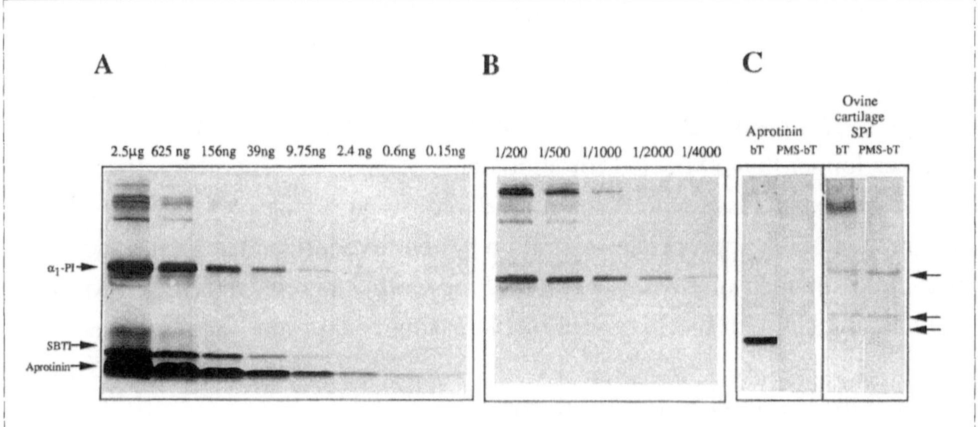

Figure 9.3 Detection of serine proteinase inhibitory proteins

Detection on Western blots using bT as probe (A, B) and verification of the binding of bT to the active site of SPIs using PMS-bT (C). In (A), 10-fold serial dilutions of a mixture of α1-PI, SBTI, and aprotinin (2.5 μg to 0.15 ng each/well) were electrophoresed on 4–20% polyacrylamide gradient gels using the Tris-glycine buffer system (22). The samples were then electroblotted to nitrocellulose, and SPIs were visualised using the bT detection system. In (B), doubling serial dilutions of normal human serum were used as samples, and SPIs on Western blots were also identified using bT. In (C), the bT was treated with PMSF to modify the active-site Ser 185 residue. The resultant PMS-bT was used as a probe alongside bT, using an ovine cartilage serine proteinase inhibitor and aprotinin as samples. Non-active-site-directed binding of several "apparent SPI species" to bT and PMS-bT was evident in the ovine SPI sample (arrows at right-hand side of figure). Aprotinin and the native ovine chondrocyte SPIs were not reactive, however, with the PMS-bT probe. Subsequent amino terminal sequencing of these 3 fragments established that they had 100% homology with concanavalin A and must have arisen from the ConcanavalinA affinity step used in the isolation of the ovine cartilage SPI (5) (modified from ref. 1 with permission from Anal. Biochem.).

9.4 Troubleshooting

General considerations

Stability of biotin succinimide esters

Solutions of biotin succinimide esters should not be stored, since they are prone to hydrolysis at neutral-alkaline pH values. At pH 7 the half-life of hydrolysis is ~ 2–4 hours; at pH 9 this drops to a few minutes. For long-term storage, biotin succinimide esters should be stored dried under dessicant below 4 °C.

Removal of free biotin from biotinylated samples

Although biotin is a small molecule (244 Da) and is relatively soluble in aqueous media (~ 0.22 mg/ml at 25 °C), its removal from proteins by dialysis or diafiltration takes longer than expected for a molecule of this size. Therefore, these steps need to be exhaustively performed, and biotinylated samples must be chromatographed on an appropriate separation system to confirm that total removal of free biotin from the sample has occurred. Free biotin can reduce the detection sensitivity attainable with biotin probes by increasing background/blank values.

Appropriate biotinylation conditions

A range of biotin incorporations into labeled probe should be undertaken, and compositional and functional assays should be used to systematically ascertain the most suitable probe to use in a particular application. Clearly, foreknowledge of the stability properties, the amino acid composition, and/or sequence of prospective serine proteinases or serine proteinase inhibitors which are to be biotinylated can aid in the formulation of appropriate labeling conditions. If lysine residues are present in the active site of the serine proteinase or serine proteinase inhibitor and are important for the biological activity of that protein, these should be protected by forming a reversible enzyme-inhibitor complex prior to biotinylation. Biotin succinimide esters are reactive primarily with the ε-amino moieties of lysine residues, particularly under alkaline conditions. Overincorporation of biotin can also lead to loss of biological activity of serine proteinases and serine proteinase inhibitory proteins;

however, biotinylation at less basic pH's can be advantageous under certain circumstances. This is why it is important to ascertain both the composition and the functional status of biotinylated probes.

Choice of biotin succinimide esters for biotinylation

A range of water-soluble and water-insoluble succinimide esters with and without spacer arms is available. In this chapter the water-soluble NHS-LC-biotin containing a 6-carbon spacer arm provided the best results of the biotin succinimide esters examined and would be the ester of choice in any future labeling procedures. The spacer arm undoubtably facilitated access of the terminal biotin moiety of the biotinylated probe to the deep biotin-binding pocket of avidin used in detection of these probes, and may also have contributed to the high sensitivity attainable with the biotinylated probes.

Storage of biotinylated samples

Biotinylated samples should be aliquoted and stored at –20 °C and repeated freeze thawings avoided to prevent breakdown of the probe and release of free biotin.

Endogenous sources of biotin

Biotin occurs as a prosthetic group in several enzymes involved in carboxylation reactions in the biosynthesis and degradation of fatty acids in bacteria, plants, fungi, and vertebrates (30–32). These enzymes include acetyl-CoA carboxylase (EC 6.4.1.2), pyruvate carboxylase (EC 6.4.1.1), propionyl-CoA carboxylase (EC 6.4.1.3), methyl-crotonyl-CoA carboxylase (EC 6.4.1.4), and geronyl-CoA carboxylase (EC 6.4.1.5). In plants, acetyl-CoA carboxylase is the only known biotin containing protein (30). All of the aforementioned biotin-containing enzymes, however, are widely distributed in bacterial, fungal, and vertebrate cells, and this should be taken into consideration if the serine proteinase or serine proteinase inhibitor sample to be examined is a crude nonfractionated cellular extract (30–32). Some of the biotin-containing enzymes are relatively large proteins, e.g., pyruvate carboxylase (M_r 600 000 tetramer in yeast, M_r 650 000 in chicken liver, M_r 620 000 in porcine liver), and thus are unlikely to be confused as SPIs in most separation systems using avidin detection systems. Monomeric fragments of the CoA carboxylases containing biotin of M_r 34–60 kDa, however, have also been identified, and these could potentially interfere in detection systems involving avidin. Samples may be pretreated with avidin to bind endogenous biotin-containing proteins prior to the detection step. Alternatively, if Western blotting is to be conducted, a small amount of avidin may be included in the blocking solution to block any endogenous source of biotin prior to addition of the avidin conjugate.

Specific considerations

Biotinylation of SLPI

Since biotin succinimide esters react primarily with the ε-amino moieties of lysine residues, incorporation of biotin into SLPI occurs readily at alkaline pH. However, SLPI is unstable at pH ≥8, and the high incorporation of biotin under these conditions can further lower the inhibitory activity of the labeled product. Conversely, SLPI is stable at pH ≤7, and if biotinylation is conducted at pH 5, it is easier to control the extent of incorporation of biotin into SLPI.

Biotinylation of trypsin

Trypsin is extremely prone to self-digestion under the alkaline conditions commonly employed for biotinylation using biotin succinimide esters (1); therefore, it is critical under all procedures to deal with trypsin in an inactive form to avoid autolysis. Trypsin is best preformed into an enzyme-inhibitor complex either with SBTI or aprotinin prior to biotinylation. Both of these inhibitors have been used successfully in our laboratory for this purpose; however, since the difference in molecular weight between trypsin and aprotinin is greater than between trypsin and SBTI, it is technically easier to separate trypsin from aprotinin than SBTI from trypsin. Trypsin is stable but inactive in glycine-HCl buffer (pH 2.0) or 3–5 mM HCl, and these solutions should be used where possible in any manipulative procedures to minimise autolytic breakdown of the biotinylated probe. Since it is essential that the biotinylated trypsin be biologically active to detect SPIs on Western blots, due consideration must be given to any storage steps of this probe in buffers where it may be enzymatically active, prior to its use as a probe in Western blotting.

Under certain circumstances some proteins which are not SPIs may be detected using the bT detection system and Western blotting (see Fig. 9.3C). One can confirm whether these represent false positive results as follows. The active-site residue of bT (Ser 185) may be modified by treatment with PMSF (Procedure 9.13) and the PMS-bT used in an analogous manner on Western blots to the bT. Any reactivity with the PMS-bT probe indicates that binding is not through the active-site Ser residue of bT, since binding of SPIs to bT is abolished by the PMSF treatment (see Fig. 9.3C). During the purification of a native ovine cartilage SPI (5) several protein species of 12–50 kDa were found to be reactive with the bT probe during Western blotting. Amino terminal sequencing of these 12–50 kDa species indicated they had a 100% homology with concanavalin A and thus represented fragments of affinity ligand from the Con A affinity chromatography step used for isolating the cartilage native SPI. These 12–50 kDa bT-reactive species were also detected using PMS-bT as a probe in Western blots. The native ovine cartilage SPI, however, was only reactive with bT, confirming its binding to the active site of the bT probe (see Fig. 9.3).

Biotinylation of PCTI-1

Potato chymotrypsin inhibitor is a tetrameric protein 39 kDa in size; this tertiary structure must be maintained in any labeling procedure with biotin. Physiological pH conditions (7.2) were found to be suitable for labeling of this protein, which could be recovered in a highly active form.

Western blotting

Blocking with 0.1% Tween 20 in TBS was found to be the most satisfactory blocking agent for nitrocellulose for the applications listed in this chapter (Table 9.1). It should be borne in mind that fat-reduced dried milk, BSA (Cohn fraction V), and serum, which are all commonly used for blocking Western blots, may potentially be a source of biotin and thus can increase the background of blots if an avidin-alkaline phosphatase conjugate is used in the visualisation procedure. If the aforementioned protein solutions must be used for blocking, a small amount of avidin may also be added to the block solution to precipitate any free biotin.

Acknowledgments

This study was partially funded by the National Health and Medical Research Council of Australia, and The Arthritis Foundation of Australia, who are gratefully acknowledged. We would also like to thank Associate Prof. P. Ghosh, Director of Raymond Purves Bone and Joint Research Laboratories, the University of Sydney, for his support and Dr R.C. Thompson, Synergen, Boulder, CO, USA, for kindly supplying SLPI.

References

1 Melrose, J., Rodgers, K. and Ghosh, P. (1994) *Anal. Biochem.* **222**, 34–43.
2 Melrose, J. and Ghosh, P. (1992) *Anal. Biochem.* **204**, 372–382.
3 Rodgers, K.J., Melrose, J. and Ghosh, P. (1995) *Anal. Biochem.* **227**, 129–134.
4 Melrose, J., Ghosh, P. and Patel, M. (1995) *Int. J. Biochem. Cell Biol.* **27**, 891–904.
5 Rodgers, K.J., Melrose, J. and Ghosh, P. (1996) *Electrophoresis.* **17**, 213–218.
6 Bayer, E.A. and Wilchek, M. (1980) *Meth. Biochem. Anal.* **26**, 1–45.
7 Billingsley, M.L., Pennypacker, K.R., Hoover, C.G. and Kincaid, R.L. (1987) *BioTechniques* **5,** 22–31.
8 Wilchek, M. and Bayer, E.A. (1989) *Trends in Biochem. Sci.***14**, 408–412.
9 Wilchek, M. and Bayer, E.A. (1988) *Anal. Biochem.* 171, 1–32.
10 Guesdon, J.-L., Ternynck and Avrameas, S. (1979) *J. Histochem. Cytochem.* **27**, 1131–1139.
11 Kay, G., Bailie, J.R., Halliday, I.M., Nelson, J. and Walker, B. (1992) *Biochem. J.* **283**, 455–459.
12 Walker, B., Cullen, B.M., Kay, G., Halliday, I.M. and McGinty, A. (1992) *Biochem. J.* **283**, 449–453.
13 Kam, C.-M., Abuelyaman, A.S., Li, Z., Hudig, D. and Powers, J.C. (1993) *Bioconjugate Chem.* **4**, 560–567.
14 Brown, A.M., George, S.M., Blume, A.J., Dushin, R.G. and Jacobsen, J.S. (1994) *Anal. Biochem.* **217**, 139–147.
15 Powers, J.C. and Kam, C.-M. (1994) *Methods Enzymol.* **244**, 442–457.
16 Green, N.M. and Toms, E.J. (1973) *Biochem. J.* **133**, 687–700.
17 Henrikson, K.P., Allen, S.H.G. and Maloy, W.L. (1979) *Anal. Biochem.* **94**, 366–370.
18 Smith, P.K., Krohn, R.I., Hermanson, G.T., Mallia, A.K., Gartner, F.H., Provenzano, M.D., Fujimoto, E.K., Goeke, N.M., Olson, B.J., Klenk, D.C. (1985) *Anal. Biochem.* **150**, 76–85.
19 Somorin, O., Tokura, S., Nishi, N. and Noguchi, J. (1978) *J. Biochem.* **85**, 157–162.
20 DelMar, E.G., Largman, C., Brodick, J.W. and Geokas, M.C. (1979) *Anal. Biochem.* **99**, 316–320.
21 Green, N.M. (1970) *Methods Enzymol.* **128**, 418–424.
22 Laemmli, U.K. (1970) *Nature* **227**, 680–685.
23 Schagger, H. and Von Jagow, G. (1987) *Anal. Biochem.* **166**, 368–379.
24 Habeeb, A.F.S.A. (1966) *Anal. Biochem.*

14, 328–336.

25 Fiedler, F., SeeMuller, U. and Fritz, H. (1984) Enzymes 3: Peptidases and Proteinases and Their Inhibitors. *In*: H.U. Bergmeyer, J. Bergmeyer and M. Grassl (eds) *Methods in Enzymatic Analysis*, 3rd Edition, Vol. 5, Verlag-Chemie, Weinheim, Germany.

26 Fritz, H. and Wunderer, G. (1983) *Arzn. Forsch. Drug Res.* **33**, 479–494.

27 Thompson, R.C. and Ohlsson, K. (1986) *Proc. Natl Acad. Sci. USA* **83**, 6692–6696.

28 Melville, J.D. and Ryan, C.A. (1972) *J. Biol. Chem.* **247**, 3445–3453.

29 Haas, G.M., Hermodson, Ryan, C.A. and Gentry, L. (1982) *Biochemistry* **21**, 752–756.

30 Nikolau, B.J., Wurtele, E.S. and Stumpf, P.K. (1984) *Plant Physiol.* **75**, 895–901.

31 Scrutton, M.C. and Fatebene, F. (1975) *Anal. Biochem.* **69**, 247–260.

32 Manchenko, G.P. (1994) Part III. Methods of detection of specific enzymes. *In*: *Handbook of Detection of Enzymes on Electrophoretic Gels*, CRC Press, Boca Raton, FL, pp 333–334.

Avidin/Biotin-Mediated Conjugation of Antibodies to Erythrocytes: An Approach for *in vivo* Immunoerythrocyte Exploration

Vladimir R. Muzykantov

Summary

Antibody-carrying red blood cells (immunoerythrocytes) may be useful as a vehicle for drug targeting and for selective elimination of antigens from the bloodstream. Such immunoerythrocytes have to circulate in the bloodstream for a prolonged time without elimination and possess high affinity to the target antigen. Streptavidin/biotin technology is useful for attachment of biotinylated antibodies (b-Ab) to biotinylated red blood cells (b-RBC). Modification of RBC with biotin hydroxysuccinimide ester provides covalent coupling of biotin residues to amino groups on the RBC membrane. b-RBC are stable in serum *in vitro* and in the bloodstream. Binding of streptavidin (SA) to b-RBC, however, leads to their lysis by complement *in vitro* and fast elimination from circulation. The mechanism of lysis is streptavidin-induced cross-linking of membrane regulators of complement. Reduction of the surface density of biotin residues on RBC membrane prevents lysis. Surface density of biotin residues on RBC may be regulated by varying biotin ester concentration during RBC biotinylation (b_n-RBC). SA specifically binds to b_n-RBC (e.g., 10^5 SA molecules per rat RBC biotinylated with 20 μM biotin, b_{20}-RBC). Modest and intermediate degrees of RBC biotinylation provide effective attachment of b-IgG to streptavidin-coated b-RBC (SA/b-RBC). SA/b_{20}-RBC bind 5×10^4 b-IgG molecules per cell. In contrast, SA attached to extensively biotinylated RBC (b_{700}-RBC) does not bind biotinylated IgG. b-Ab/SA/b_{20}-RBC are stable in serum and bind to immobilized antigen *in vitro*. ^{51}Cr-labeled serum-stable b-Ab/SA/b_{20}-RBC circulate without marked elimination and/or lysis in the bloodstream after iv. injection in rats. Biodistribution of b-Ab/SA/b_{20}-RBC in rats is similar to that of control RBC. Therefore, b-Ab/SA/b_{20}-RBC satisfy major requirements of a vehicle for drug targeting (i.e., high affinity to the target and biocompatibility) and can be explored *in vivo*.

10.1 Introduction

Red blood cells (RBC) are an appropriate vehicle for targeting drugs (1). RBC are a nonimmunogenic, available, and physiologic vehicle. Both large inner volume and large surface of the plasma membrane of RBC could be used for loading or coupling of a drug. To provide the targeting, an antibody against the target antigen may be attached to RBC. Such antibody-carrying RBC (immunoerythrocytes) may also be useful for selective clearance of circulating pathogens and harmful antigens from the bloodstream (2). Thus, technology of an antibody attachment to RBC is important for drug targeting and blood clearance.

This technology must meet at least 2 major requirements. First, immunoerythrocytes should possess affinity to the target. Second, immunoerythrocytes should be stable in serum and circulate in the bloodstream for a prolonged time, because lysis and elimination of immunoerythrocytes will minimize targeting.

Streptavidin-mediated cross-linking of biotinylated molecules (streptavidin/biotin technology, ref. 3) offers an approach for attachment of biotinylated antibodies (b-Ab) to biotinylated RBC (b-RBC). Streptavidin attached to b-RBC (SA/b-RBC) may serve as a universal binding site for biotinylated antibodies, enzymes, cytokines, and so on (4, 5).

Biotinylation of RBC was developed almost 2 decades ago for *in vitro* investigation of RBC membrane (6). Recently, the use of b-RBC was explored *in vivo*, in laboratory animals (7) and in humans (8), as a new approach for tracing circulating RBC in the bloodstream. Several studies have demonstrated that biotinylation does not significantly alter biocompatibility of RBC (7–9).

However, polyvalent binding of avidin (or SA) to b-RBC induces their lysis by complement (10). The mechanism of lysis is streptavidin-induced cross-linking of biotinylated regulators of complement in b-RBC membrane, leading to their inactivation (10, 11). In contrast, monovalent attachment of SA to RBC, which does not lead to cross-linking of the membrane components, does not induce lysis (12). Reduction of the surface density of biotin residues on the membrane of b-RBC is a useful approach to decrease the valency of SA binding to b-RBC and to obtain serum-stable immunoerythrocytes (13). In this chapter I describe the methodology of preparation of serum-stable immunoerythrocytes (b-Ab/SA/b-RBC) and their properties studied *in vitro* and *in vivo*.

10.2 Technical Procedures

Materials

Streptavidin, 6-biotinylaminocaproic acid *N*-hydroxysuccinimide ester (long-arm biotin ester, BxNHS) and polyclonal goat antibody against mouse IgG were purchased from Calbiochem (San Diego, CA). Iodogen was obtained from Pierce (Rockford, IL). Both (^{51}Cr)Cl and Na(^{125}I) isotopes were from Amersham (Arlington Heights, IL). Normal mouse IgG, bovine serum albumin (BSA), dimethylformamide (DMF), and components of buffers were from Sigma (St. Louis, MO). Proteins were radiolabeled with ^{125}I by the Iodogen method according to the manufacturer's recommendations. Goat antibody against mouse IgG, as well as nonimmune mouse IgG, was biotinylated with BxNHS at biotin/IgG molar ratio 10:1 in the reaction mixture, as described (14). Gelatin-veronal buffer (GVB) is a water solution containing 3 mM diethylbarbituric acid, 1 mM sodium salt of diethylbarbituric acid, 145 mM NaCl, 1.8 mM $MgCl_2$, 0.25 mM $CaCl_2$, 0.1% gelatin, pH 7.4.

Sprague-Dawley male rats (Charles River Breeding Laboratories, Kingston, NY) weighing 200–300 g, were anesthetized with i.p. injection of sodium pentobarbita (50 mg/kg), and the peritoneal cavity was opened. To obtain serum, nonheparinized blood was collected from the peritoneal cavity after transection of the descending aorta. After 2 hours incubation at 4 °C, serum was separated by centrifugation. To obtain RBC, blood from the peritoneal cavity of anesthetized rats was collected in heparin. Human blood was obtained from healthy volunteers by venapuncture.

Procedure 10.1

Modification of RBC with biotin/streptavidin and attachment of biotinylated IgG

General procedure of RBC biotinylation with BxNHS has been described previously (4, 13). Stock solution of BxNHS in DMF (100 mM BxNHS in anhydrous DMF) may be prepared in advance (it will keep for several weeks at –20 °C). Centrifuge 2 ml of fresh heparinized blood at 1500 rpm for 5 min and eliminate supernatant (plasma). Resuspend and wash the pellet (100% RBC suspension) with phosphate-buffered saline (PBS) by standard centrifugation (10 µl of PBS per 1 ml of pellet, 1500 rpm, 5 min, 4 times). Add 0.9 ml of PBS to 0.1 ml of RBC pellet (i.e., make 10% suspension of washed RBC). Add 100 µl of 300 mM boric acid (pH 9.0) to 1.0 ml of 10% RBC. Then add BxNHS in DMF to this suspension to obtain the final BxNHS concentration in the reaction mixture within the range 2–3000 µM. When varying BxNHS addition, keep addition of the vehicle, DMF at standard levels (10 µl DMF per 1 ml of 10% RBC suspension). For example, to obtain b_{20}-RBC, at first add 2 µl of stock solution of 0.1 M BxNHS/DMF to 98 µl DMF. Then add 10 µl of this fresh 2 mM BxNHS/DMF to 1 ml of 10% RBC and mix well. After 30 min incubation (periodic gentle shaking at 20 °C), eliminate excess of nonreacted BxNHS from the reaction mixture by standard centrifugation with PBS containing 2 mg/ml of BSA (BSA-PBS). Prepare 10% suspension of biotinylated RBC in BSA-PBS. In the following text, micromolar concentration of BxNHS in the reaction mixture will be defined with a b_n-RBC abbreviation, that is, b_{20}-RBC means RBC biotinylated at 20 µM BxNHS.

To attach streptavidin to b_n-RBC, add 20 µl of SA stock solution (1 mg/ml in PBS) to 100 µl of a 10% suspension of b_n-RBC and mix well. This provides addition of 1 µg of SA per 5×10^6 b_n-RBC (about 2×10^6 SA molecules per b_n-RBC). After 30 min incubation (periodic gentle shaking, 20 °C), remove nonbound SA by standard centrifugation in BSA-PBS. To quantitate SA attachment to b_n-RBC, add radiolabeled SA to b_n-RBC and measure radioactivity in the pellet after washing off nonbound SA.

To attach biotinylated antibody, as well as control IgG (b-Ab and b-IgG) to SA/b-RBC, add 5 µl of stock solution b-Ab or b-IgG (1 mg/ml in PBS) to 100 µl of a 10% suspension of SA/b_n-RBC and mix well. This provides addition of 1 μg of b-Ab or b-IgG per 2×10^7 SA/b_n-RBC (about 1.5×10^5 molecules per SA/b_n-RBC). Incubate b-Ab or b-IgG with a 10% suspension of SA/b-RBC for 1 hour (periodic gentle shaking, 20 °C). Remove nonbound b-IgG or b-Ab by standard centrifugation with BSA-PBS. To quantitate binding of b-IgG to SA/b-RBC, use radiolabeled b-IgG as a tracer.

Procedure 10.2

Estimation of lysis of immunoerythrocytes in homologous serum

Lysis of RBC, b-RBC, SA/b-RBC, and b-Ab/SA/b-RBC by fresh homologous serum can be studied in 96-well microtest plates, estimating light transmission of RBC suspensions as described earlier (15). Use gelatin-veronal buffer containing Ca^{2+} and Mg^{2+}(GVB) as a diluent for serum and RBC. Prepare a 2% suspension of RBC in GVB by addition of 20 μl of a 10% RBC suspension to 80 μl of GVB. Add 50 μl of serum (1/2–1/20 dilutions in GVB) and 50 μl of a 2% suspension of RBC preparations in wells of the microtest plate, and incubate for 1 hour at 37 °C. Then measure lysis by reading at 630 nm in an ELISA reader. Use wells containing water instead of GVB as a standard for 100% lysis. Use wells containing serum-free GVB as a standard for zero lysis. Calculate percentage of lysis by the formula: lysis % = [(A0%–A)/(A0%–A100%)] × 100, where A100% is the absorbance in the wells with water, A0% is the absorbance in the wells with serum-free GVB, and A is the absorbance.

Procedure 10.3

Estimation of immunoerythrocytes binding to antigen

To study the binding of soluble antigen to immunoerythrocytes, add ^{125}I-labeled antigen to suspension of immunoerythrocytes in BSA-PBS and incubate the mixture for 1 hour at 20 °C (the procedure is described in detail in ref. 13). As a negative control use ^{125}I-labeled BSA instead of antigen or normal RBC instead of immunoerythrocytes. Eliminate nonbound radioactivity by standard centrifugation and estimate antigen binding to RBC by measuring radioactivity in the pellet. To study binding of immunoerythrocytes to the target, immobilize an antigen on the bottom of plastic wells of 24 well culture plates, as described (4). In the present study, mouse IgG was used as a model antigen. Add 1 µg of mouse IgG per well in 0.5 ml of PBS. After overnight incubation at 4 °C, eliminate nonbound antigen by washing with water. Block sites of nonspecific binding in wells by 1-hour incubation with 1 ml of BSA-PBS. Then add 10 µl of a 10% suspension of immunoerythrocytes (b-anti-IgG/SA/b-RBC, 10^7 cells/well) per well containing 0.5 ml of BSA-PBS and mix well. After 2 hours incubation at 20 °C, eliminate nonbound RBC by washing with PBS. To quantitate RBC binding to the antigen, lyse attached RBC by addition of 1 ml of water per well. Estimate the amount of bound RBC in wells by measurement of hemoglobin absorbance in lysates at 405 nm.

Procedure 10.4

Estimation of the fate of immunoerythrocytes after iv. injection in rats

To trace immunoerythrocytes after *in vivo* administration, add ^{51}Cr isotope (10 µCi per 100 µl of a 10% RBC suspension) to SA/b_{20}-RBC suspension simultaneously with b-Ab and incubate for 1 hour at 20 °C. Eliminate excess of isotope by standard centrifugation. This procedure provides 35–50% effectiveness of RBC radiolabeling. Inject 20–50 µl of a 10% suspension of ^{51}Cr-labeled b-Ab/SA/b_{20}-RBC via the tail vein in anesthetized rats (see Troubleshooting). At the indicated time after injection, anesthetized rats should be sacrificed by exsanguination. Collect blood and internal organs. Rinse organs with saline until free of blood, weigh organs and an aliquot of blood, and measure radioactivity of ^{51}Cr in blood and internal organs using a gamma counter. After this, separate plasma by centrifugation of blood and determine radioactivity in plasma. Calculate results as cpm per gram of tissue, blood, or plasma. Lysis of immunoerythrocytes in the bloodstream may be calculated by the formula: lysis % = (cpm per g of plasma/cpm per g of blood) × 50.

10.3 Results and Discussion

Modification of RBC with biotin/streptavidin and attachment of b-IgG to RBC

RBC biotinylation is a simple procedure that provides sites for high-affinity binding of streptavidin on the RBC membrane. Several different biotin derivatives are available which allow the coupling of biotin residue to various components of RBC membrane: sugar moieties of glycoproteins and glycolipids (16), sialic acid (17), SH-groups (18), tyrosine residues (19). The most useful, however, are succinimide esters of biotin (BNHS), which provide covalent coupling of biotin to amino groups of RBC membrane (4, 13, 20). There are several succinimide esters of biotin: short ester (BNHS), long ester, possessing a 6-Å spacer between biotin and the reactive group (BxNHS), and very long ester (BxxNHS, possesses a 12–Å spacer). Long ester is preferable for immunoerythrocyte preparation because the spacer provides additional steric freedom for streptavidin and biotinylated antibody attached to b-RBC and thus enhances the probability of effective interaction of RBC-bound antibody with its antigen.

Figure 10.1 shows binding of radiolabeled streptavidin to b_n-RBC. Rat RBC have been biotinylated in the range of BxNHS concentrations in the reaction mixture equal to 2–700 μM (b_2–b_{700}RBC). Human RBC have been biotinylated in the range 3–3000 μM BxNHS (b_3–b_{3000}RBC). Maximal SA attachment was 5×10^5 molecules per rat b_{700}-RBC and 10^6 molecules per human b_{3000}-RBC. Both rat B_{20}-RBC (i.e., RBC biotinylated at 20 μM BxNHS) and human b_{30}-RBC bind about 10^5 SA molecules per b-RBC. Thus, increase in biotin ester concentration during biotinylation of RBC provides more binding sites for SA on the RBC membrane, with a saturation level close to 10^6 SA molecules/RBC.

Radiolabeled b-IgG effectively binds to b-RBC coated with streptavidin (SA/b-RBC). Importantly, streptavidin attached to RBC biotinylated at a high concentration of BxNHS possesses reduced ability to bind b-IgG (Fig. 10.2). Thus, attachment of b-IgG to rat SA/b_{700}-RBC was dramatically lower than attachment of b-IgG to SA/b_{70}-RBC, whereas b_{700}-RBC possess more SA bound per cell (see Fig. 10.1A). Thus, biotin residues coupled to the membrane of b_{700}-RBC occupy most of the biotin-binding sites of SA and block the subsequent attachment of b-IgG (13, 21). Therefore, to obtain effective attachment of b-IgG or b-Ab to SA/b-RBC, the

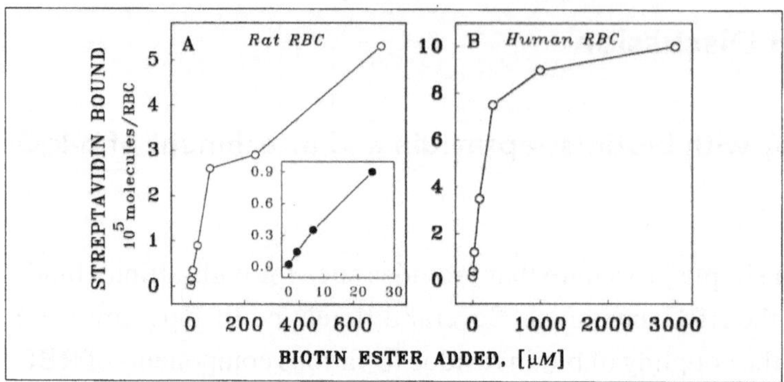

Figure 10.1 Binding of radiolabeled streptavidin to biotinylated rat (A) and human (B) red blood cells

Freshly washed RBC (10% suspension in saline) were biotinylated with BxNHS (6-biotinylaminocaproic acid N-hydroxysuccinimide ester) at concentrations indicated on the abscissa as described in Technical Procedures (30 min incubation with BxNHS at pH 9.0). Biotinylated RBC were incubated for 30 min with radiolabeled streptavidin (20 μg of streptavidin per 100 μl of a 10% suspension of biotinylated RBC). After washing off nonbound streptavidin, radioactivity associated with RBC was determined. Insert in (A) shows binding of streptavidin to rat RBC biotinylated at modest levels. Results are for a representative experiment.

optimal extent of RBC biotinylation should be determined in order to avoid total occupation of biotin-binding sites of SA by RBC-coupled biotin residues. In our studies utilizing rat and human RBC, incubation of a 10% RBC suspension with 20–300 μM BxNHS provides the optimal extent of RBC biotinylation in terms of streptavidin-mediated attachment of b-IgG. Figure 10.3 demonstrates saturable and effective binding of radiolabeled b-IgG to SA/b_{20}-RBC (rat) and to SA/b_{100}-RBC (human). As high as 50% of the added b-IgG binds to SA/b-RBC. Quantitatively, about 5×10^4 molecules of b-IgG bind per rat SA/b_{20}-RBC and 7×10^4 molecules of b-IgG bind per human SA/b_{100}-RBC.

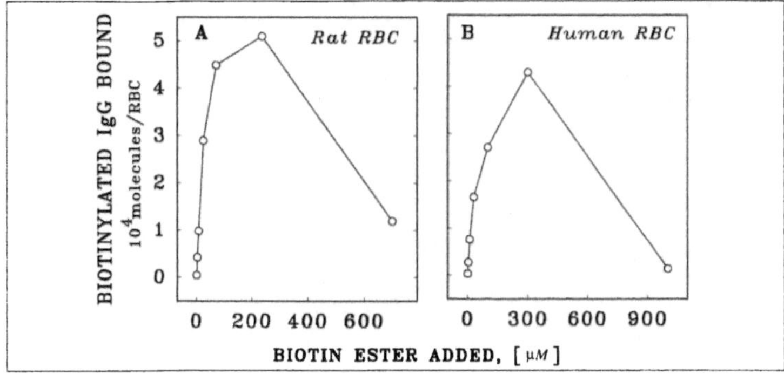

Figure 10.2 Attachment of radiolabeled biotinylated mouse IgG to rat (A) and human (B) SA/bn-RBC

RBC were biotinylated at indicated micromolar concentrations of BxNHS, and streptavidin (SA) was bound to b_n-RBC as described in Technical Procedures. Streptavidin-coated b_n-RBC (SA/b_n-RBC) were incubated with radiolabeled biotinylated mouse IgG (1.5 × 10^5 b-IgG molecules per SA/b_n-RBC, 1 hour, room temperature). After washing off nonbound radiolabeled b-IgG, radioactivity associated with RBC was determined. Note reduced b-IgG-binding capacity of streptavidin-coated extensively biotinylated RBC (rat SA/b_{700}-RBC and human SA/b_{1000}-RBC), despite the fact that extensively biotinylated RBC bind maximal amount of streptavidin (see Fig. 10.1). This result indicates that most of streptavidin molecules are attached to rat b_{700}-RBC and human b_{1000}-RBC in a polyvalent fashion and, therefore, possess no free binding sites for biotin. Results are for a representative experiment.

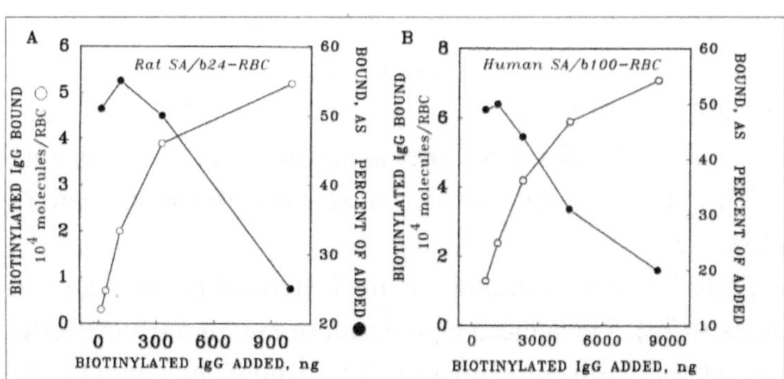

Figure 10.3 Attachment of radiolabeled biotinylated mouse IgG to rat SA/b24-RBC (A) and human SA/b100-RBC (B)

Binding of radiolabeled b-IgG to SA/b_{24}-RBC and SA/b_{100}-RBC has been performed as described in Technical Procedures and in the legend to Figure 10.2. Open circles show attachment of b IgG to SA/b-RBC, achieving 5 × 10^4–7 × 10^4 b-IgG molecules per RBC at maximal addition of b-IgG. Closed circles show effectiveness of the attachment, expressed as percent of added b-IgG. Under nonsaturating conditions, about 50% of added b-IgG binds to both types of streptavidin-coated biotinylated RBC (SA/b_{24}-RBC and SA/b_{100}-RBC). Results are for a representative experiment.

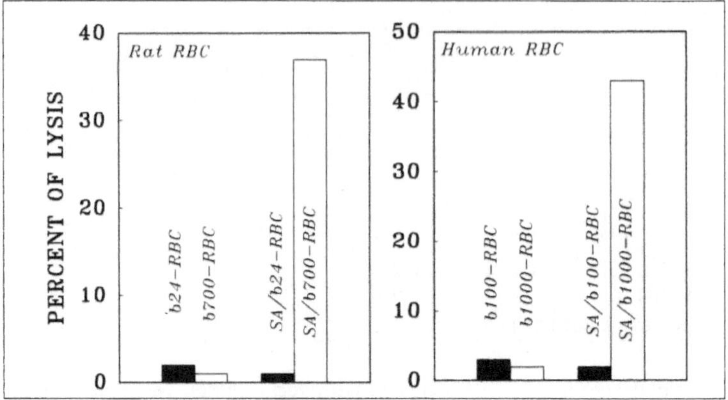

Figure 10.4 Lysis of biotinylated RBC and streptavidin-coated biotinylated RBC in normal fresh autologous serum

Biotinylated RBC (rat b_{24}-RBC and b_{700}-RBC, human b_{100}-RBC and b_{1000}-RBC) or streptavidin-coated biotinylated RBC (rat SA/b_{24}-RBC and SA/b_{700}-RBC, human SA/b_{100}-RBC and SA/b_{1000}-RBC) were incubated 1 hour at 37 °C with fresh homologous serum at final dilution 1/5. Lysis of RBC has been estimated by measurement of light transmission of suspensions, as described in Technical Procedures. Note lysis of streptavidin-coated extensively biotinylated RBC. Results are for a representative experiment.

Stability of immunoerythrocytes in serum: *in vitro* study

Several groups have reported that b-RBC circulate in the bloodstream similar to nonmodified RBC, without marked elimination or lysis after systemic administration in various animal species (7–9). However, previous studies demonstrate that attachment of avidin or SA to b-RBC reduces biocompatibility of b-RBC, mainly by 2 mechanisms (10–13).

First, avidin-carrying RBC bind to homologous nucleated cells, for example, Kupffer cells and fibroblasts (22). The adhesive potential of avidin-carrying RBC does not depend on the mode of avidin attachment to RBC and is mediated by the adhesive potential of avidin. Due to its strong positive charge, avidin binds nonspecifically to various negatively charged cells, particles, and molecules (23). In contrast, SA is a neutral sugar-free protein (23); SA-carrying RBC do not bind to nucleated cells (22).

Second, both avidin and SA induce lysis of b-RBC by autologous serum (10, 13). Figure 10.4 shows that RBC are stable in serum regardless of the extent of biotinylation. Attachment of SA induces lysis of rat b_{700}-RBC and human b_{1000}-RBC.

Noteworthy, SA attached to both these types of b_n-RBC does not bind b-IgG (see Fig. 10.2). In contrast, rat b_{24}-RBC and human b_{100}-RBC are stable in serum after attachment of SA. Moreover, attachment of b-IgG or b-Ab to rat SA/b_{24}-RBC and to human SA/b_{100}-RBC does not induce their lysis in serum, even in the presence of antigen (13, 16). Thus, reduction of the surface density of biotin residues on the RBC membrane leads to monovalent binding of SA to b-RBC and provides complement-stable immunoerythrocytes.

Binding of immunoerythrocytes to antigen: *in vitro* study

High affinity to the target antigen is an important requirement for immunoerythrocytes; this affinity should provide their effective binding to target cells, tissues, particles, or molecules. The simple way to evaluate the affinity properties of immunoerythrocytes is an assessement of binding of radiolabeled antigen to immunoerythrocytes in suspension. Figure 10.5 shows specific and saturable binding of radiolabeled antigen to human b-Ab/SA/b_{100}-RBC. In terms of preparation of immunoerythrocytes for selective elimination of circulating antigens from the bloodstream, this assay may also serve as a functional test (13).

Targeting of RBC-loaded drug to the target cell or tissue requires binding of immunoerythrocytes to the antigen-exposed surface rather than to soluble antigen. However, interaction of antibody-carrying RBC with immobilized antigen may be

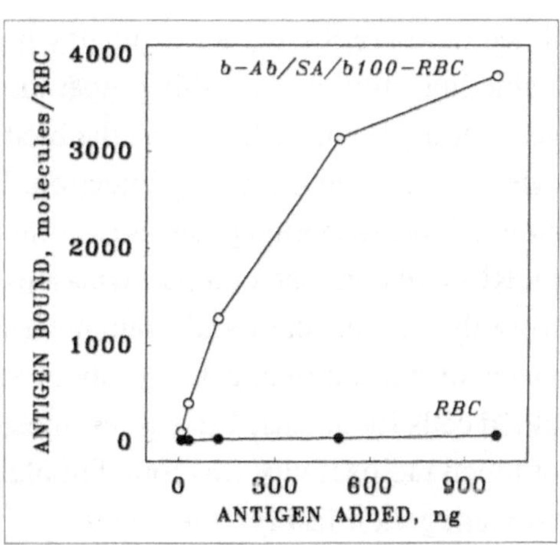

Figure 10.5 Binding of soluble radiolabeled antigen (human IgM) to human SA/b_{100}-RBC carrying biotinylated mouse monoclonal antibody against human IgM

Serum-stable biotinylated human RBC coated with streptavidin and biotinylated monoclonal antibody against human IgM (b-mAb/SA/b_{100}-RBC) were incubated with indicated amounts of radiolabeled human IgG for 1 hour at 20 °C. After washing off nonbound IgM, radioactivity associated with RBC was determined. Binding of antigen to suspension of unmodified human RBC (closed circles) is shown as a control. Results are for a representative experiment.

restricted by steric limitations imposed by (i) orientation of antibody and antigen molecules on the RBC membrane and on the target surface, and (ii) surface densities of antibody and antigen. Therefore, affinity of immunoerythrocytes in terms of drug targeting should be tested using immobilized antigen. Table 10.1 shows the result of such a test for serum-stable rat immunoerythrocytes (b-Ab/SA/b_{24}-RBC) and documents that b-Ab/SA/b_{24}-RBC, but not b-IgG/SA/b_{24}-RBC, bind specifically to the antigen-coated surface.

Table 10.1 Binding of immunoerythrocytes to an antigen-coated plastic surface

	Antigen	BSA
b-Ab/SA/b_{24}-RBC	$2.9 \times 10^6 \pm 10^5$	$2 \times 10^4 \pm 10^4$
b-IgG/SA/b_{24}-RBC	$3 \times 10^4 \pm 10^4$	$4 \times 10^4 \pm 10^4$

Plastic wells were coated with antigen (mouse IgG) or with albumin as described in Technical Procedures. Serum-stable streptavidin-coated biotinylated rat RBC carrying biotinylated goat antibody against mouse IgG (b-Ab/SA/b_{24}-RBC) or control nonspecific biotinylated IgG (b-IgG/SA/b_{24}-RBC) were incubated in antigen-coated wells and in BSA-coated wells as described in Technical Procedures. After washing off nonbound RBC, the amount of bound RBC was determined by measurement of hemoglobin in the wells, as described in Technical Procedures. The data are presented as

Biodistribution and circulation of serum-stable immunoerythrocytes in rats

Since rat b-Ab/SA/b_{24}-RBC display high affinity to immobilized antigen and are stable in serum *in vitro*, this preparation might be also appropriate for *in vivo* investigation. Figure 10.6 shows the biodistribution of ^{51}Cr-labeled immunoerythrocytes in rats 1 hour after iv. injection. Both b_{24}-RBC and b-Ab/SA/b_{24}-RBC have similar biodistribution patterns as compared with control nonmodified ^{51}Cr-labeled rat RBC. In contrast, complement-sensitive SA/b_{700}-RBC display fast elimination from the bloodstream and dramatic enhancement of liver uptake. Based on estimates of plasma radioactivity, about 98% of injected radiolabel circulates in the blood cells for at least 1 day after injection of b-Ab/SA/b_{24}-RBC. In contrast, 75% of blood radioactivity was found in plasma 1 hour after injection of SA/b_{700}-RBC, indicating their fast lysis *in vivo*.

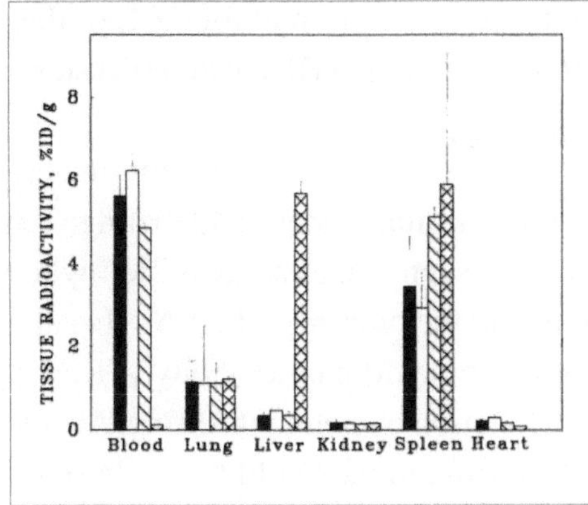

Figure 10.6 Biodistribution of 51Cr-labeled immunoerythrocytes in rats 1 hour after iv. injection

Radiolabeled RBC were injected in the tail vein of normal rats. The following preparations were injected: control RBC (closed bars); b_{24}-RBC (open bars); b-Ab/SA/b_{24}-RBC (hatched bars) and SA/b_{700}-RBC (crossed bars). The data are presented as a percent of injected radioactivity per gram of tissue, $M \pm SD$, $n = 3$. Note rapid elimination of SA/b_{700}-RBC from the blood and accumulation of SA/b_{700}-RBC-associated radioactivity in the liver. Other preparations circulate for a prolonged time.

Conclusion

Rat immunoerythrocytes b-Ab/SA/b_{24}-RBC are stable in serum *in vitro*, possess high affinity to the target antigen, and have normal biodistribution and prolonged life span *in vivo*. Therefore, the use of b-Ab/SA/b_{24}-RBC may be explored as a new device for drug targeting and elimination of circulating antigens.

10.4 Troubleshooting

1. Presence of extraneous protein(s) in RBC suspension dramatically reduces RBC biotinylation and induces irreproducibility in the extent of RBC biotinylation, because amino groups of soluble proteins are suitable targets for biotin succinimide esters. Therefore, complete elimination of plasma proteins by centrifugation is very important in order to obtain reproducible biotinylation of RBC. For the same reason, BSA or other carrier protein should be omitted until biotinylation is complete.
2. Avoid lysis of RBC during manipulations. First, RBC lysis during (or before) biotinylation will reduce biotinylation, since hemoglobin released from RBC will consume biotin ester. Second, lysed or damaged RBC undergo fast elimination after injection in animals (see below). There are several possible reasons for RBC lysis: (i) mechanical damage of RBC by manipulations; (ii) hypoton-

ic stress (osmotic pressure of buffer solution should be optimal for RBC of the given animal species; 150 mM NaCl is isotonic for human RBC); and (iii) storage of RBC *in vitro*.

3. The reticuloendothelial system (RES) effectively recognizes and eliminates RBC damaged by manipulations during antibody attachment and labeling. To protect RBC, add carrier protein (BSA) to the suspension as soon as biotinylation is completed, and carry out all procedures in the presence of BSA. This also allows reduction of nonspecific binding of streptavidin or antibody to RBC. Avoid vortexing; resuspend RBC after centrifugation by gentle pipetting. Avoid generating air bubbles in the suspension. Centrifugation should be as short as possible; 5 min is enough to sediment RBC from a 10 ml suspension at 1500 rpm. Use fresh immunoerythrocytes for injection in animals.
4. Streptavidin may induce aggregation of b-RBC (even biotinylated at low extent), when a nonsaturating concentration of streptavidin is added to b-RBC. Aggregated SA/b-RBC cannot be used for *in vivo* study. For example, both lung and RES will take up aggregated RBC due to mechanical retention of aggregates in the microvasculature of these organs. To avoid aggregation, always add saturating amounts of SA to the b-RBC suspension (such addition is described in the Technical Procedures). Nonsaturating addition of b-Ab or b-IgG also may induce aggregation of SA/b-RBC. To avoid addition of large amounts of b-Ab, which may be expensive, use monobiotinylated antibodies, which will not induce aggregation of SA/b-RBC even under nonsaturating conditions. A method for evaluating antibody biotinylation has been described (14).
5. Do not use more than a 2% suspension of RBC in the hemolytic assay utilizing 96-well microtest plates described above (for details see ref. 15). The major advantage of this assay based on the measurement of light transmission through the RBC suspension is an opportunity to compare a large number of samples under standard conditions (e.g., various RBC preparations, various serum dilutions, etc.). However, sensitivity and accuracy of this assay are lower than the measurement of hemoglobin release from RBC. For the latter assay, incubate RBC with serum at 37 °C, centrifuge the suspension at 1500 rpm for 5 min, and measure absorbance of the supernatants at 405 nm. Use RBC-free serum as a blank at A405 nm.
6. Binding of immunoerythrocytes to immobilized antigen should be studied in 24-well culture plates, not in 96-well microtest plates. Washing out of nonbound RBC in the small wells is inaccurate and irreproducible. On the other hand, im-

munoerythrocyte binding to immobilized antigen dramatically depends on the antigen surface density (21). Binding of an antigen to plastic varies, depending on (i) the kind of plastic; (ii) the antigen; and (iii) the buffer solution. For example, about 300 ng of IgG v. 30–50 ng of streptavidin binds to plastic wells after overnight incubation of 1 μg of protein in PBS. To estimate binding of antigen to plastic, immobilize radiolabeled antigen using standard procedures. After washing out the nonbound antigen and incubating for 1 hour with BSA-PBS, add 1 ml of boiling SDS solution per well and measure radioactivity in eluates.

7. Animal studies. To reduce loss of ^{51}Cr-labeled b-Ab/SA/b_{20}-RBC in the syringe and increase the reproducibility of injection, dilute ^{51}Cr-labeled b-Ab/SA/b_{20}-RBC with unmodified washed RBC by addition of 50 μl of a 10% suspension of ^{51}Cr-labeled b-Ab/SA/b_{20}-RBC to 0.95 ml of a 10% suspension of unmodified washed RBC, and inject 0.2–0.5 ml of this mixture per rat. Do not take blood samples from the tail blood vessels after injection of ^{51}Cr-labeled RBC into the tail vein.

Acknowledgments

The author thanks associates and collaborators who have contributed to previously published studies used as a basis for writing of this chapter: Drs M. Smirnov, G. Samokhin, A. Klibanov, S. Domogatsky, and D. Sakharov (Cardiology Research Center, Moscow, Russia); Dr R. Taylor (University of Virginia, Charlottesville, VA); Drs A. Zaltsman, B.P. Morgan, and C.V. den Berg (University of Wales, Cardiff, UK), Drs E. Atochina and A.B. Fisher (University of Pennsylvania, Philadelphia, PA), and Drs J.C. Murciano and A. Herraez (University of Alcala, Alcala de Henares, Spain). The author thanks Dr Aron B. Fisher for reading the manuscript and for making valuable remarks.

References

1 Poznansky, M. and Juliano, R. (1984) *Pharmacol. Rev.* **36**, 277–335.

2 Taylor, R., Reist, C., Sutherland, W., Otto, A., Labuguen, R. and Wright, E. (1992) *J. Immunol.* **148**, 2462–2468.

3 Wilchek, M. and Bayer, E. (1988) *Anal. Biochem.* **174**, 1–32.

4 Samokhin, G., Smirnov, M., Muzykantov, V., Domogatsky, S. and Smirnov, V. (1983) *FEBS Lett.* **154**, 257–261.

5 Magnani, M., Chiarantini, L., Vittoria, E., Mancini, U., Rossi, L. and Fazi, A. (1992) *Biotech. Appl. Biochem.* **16**, 188–194.

6 Orr, G. (1981) *J. Biol. Chem.* **256**, 761–766.

7 Suzuki, T. and Dale, G. (1987) *Blood* **70**, 791–795.

8 Cavill, I., Trevett, D., Fisher, J. and Hoy, T. (1988) *Br. J. Haematol.* **70**, 491–493.

9 Muzykantov, V., Seregina, N. and Smirnov, M. (1992) *Int. J. Art. Org.* **15**, 620–627.

10 Muzykantov, V., Smirnov, M. and Samokhin, G. (1991) *Blood* **78**, 2611–2618.

11 Zaltsman, A., van den Berg, C., Muzykantov, V. and Morgan, B. (1995) *Biochem. J.* **301**, 651–656.

12 Muzykantov, V., Smirnov, M. and Samokhin, G. (1992) *Biochim. Biophys. Acta* **1107**, 119–125.

13 Muzykantov, V. and Taylor, R. (1994) *Anal. Biochem.* **223**, 142–148.

14 Muzykantov, V., Gavriluk, V., Reinecke, A., Atochina, E., Kuo, A., Barnathan, E. and Fisher, A. (1995) *Anal. Biochem.* **226**, 279–287.

15 Muzykantov, V., Samokhin, G., Smirnov, M. and Domogatsky, S. (1985) *J. Appl. Biochem.* **7**, 223–227.

16 Roffman, E., Meromsky, L., Ben-Hur, H., Bayer, E. and Wilchek, M. (1986) *Biochim. Biophys. Res. Commun.* **136**, 80–85.

17 Bayer, E., Ben-Hur, H. and Wilchek, M. (1988) *Anal. Biochem.* **170**, 271–281.

18 Bayer, E., Safars, M. and Wilchek, M. (1987) *Anal. Biochem.* **161**, 262–271.

19 Wilchek, M., Ben-Hur, H. and Bayer, E. (1986) *Biochim. Biophys. Res. Commun.* **138**, 872–879.

20 Simpson, G., Born, J. and Gain, G. (1980) *Mol. Biochem. Parasitol.* **4**, 243–253.

21 Muzykantov, V. and Murciano, C. (1996) Biotechnol. *Appl. Biochem; in press*

22 Muzykantov, V., Zaltsman, A., Fuki, I., Smirnov, M., Samokhin, G. and Romanov, Yu. (1993) *Biochim. Biophys. Acta* **1179**, 148–156.

23 Duhamel, R. and Whitehead, J. (1990) *Methods Enzymol.* **184**, 201–207.

Biotin *in vitro* Translation: A Nonradioactive Method for the Synthesis of Biotin Labeled Proteins in a Cell-Free System

Hans-Joachim Hoeltke, Irene Ettl, Edith Strobel, Hermann Leying, Maria Zimmermann, and Richard Zimmermann

Summary

In vitro translation of mRNAs into proteins using, for example, reticulocyte lysates or wheat germ extracts is frequently applied to study the coding capacity of RNAs or cDNAs and the functional effects of mutations. Usually, a natural and purified RNA or an mRNA transcript synthesized from cloned cDNA *in vitro* with a phage RNA polymerase (SP6, T3, or T7) is added to an *in vitro* translation system in order to program the synthesis of the encoded protein. *In vitro* translation assays are traditionally monitored by following the incorporation of radiolabeled [^{35}S]methionine into newly synthesized protein. We have optimized an alternative nonradioactive *in vitro* translation method that labels newly synthesized proteins with biotin. $tRNA^{Lys}$ is first aminoacylated with lysine. The lysine moiety is then chemically labeled with biotin at the ε-NH_2-group. When biotin-lysine-$tRNA^{Lys}$ is added to translation systems, the biotinylated lysine is incorporated into the growing polypeptide chain. After electrophoresis and transfer of the translation products to PVDF or nitrocellulose membranes, the biotin-labeled proteins are detected with streptavidin coupled to horseradish peroxidase and a chemiluminescent reaction of the marker enzyme with luminol/iodophenol. The chemiluminescent signals are recorded by a 0.1 to 10-minute exposure of the blot to X-ray film. For the translation of viral and *in vitro* transcribed RNAs, this nonradioactive method yields equivalent results in comparison to the radioactive method. In addition, biotin-labeled translation products are biologically functional: biotinylated precursor proteins are transported and processed correctly by dog pancreas microsomes; transcription factors synthesized by biotin *in vitro* translation bind specifically to their DNA recognition sequences; and biotin-modified luciferase keeps its enzymatic activity. The major advantage of the biotin *in vitro* translation system is that no radioactivity is required, and the method is easy, economical, reproducible, and fast – the whole nonradioactive procedure, from translation to detection, can be completed within 6 hours.

11.1 Introduction

In vitro translation is a classical method of molecular biology (2). Natural RNAs or RNAs synthesized *in vitro* (4, 16) are added to translation systems and program the synthesis of specific encoded proteins. Extracts have been derived from different sources for this purpose: (i) rabbit reticulocyte lysates (12, 20), (ii) wheat germ extracts (7, 23), (iii) yeast extracts (27), and (iv) bacterial extracts from *Escherichia coli* (6, 22, 30) or, for example, *Staphylococcus aureus*. In all these assays protein synthesis is typically monitored by the incorporation of a radioactively labeled amino acid residue, usually [^{35}S]methionine or [^{3}H]leucine. Radioactively labeled proteins are subjected to SDS-polyacrylamide gel electrophoresis (PAGE) and subsequently detected by autoradiography on X-ray films, with phosphoimaging devices, or by scintillation counting. The use of lysine-tRNA^{Lys}, modified at the ε-NH_2-group of lysine (13, 14) with cross-linking agents (8–11, 14, 15, 17, 18, 29), fluorescent tags (5), or biotin (19) for the *in vitro* synthesis of nonradioactively modified proteins has been described earlier. Although these reports provided evidence for the utility of alternative nonradioactive *in vitro* translation methods, they have not found very wide application to date. We have further optimized the components and the protocols of the complete procedure with respect to sensitivity, speed, reproducibility, and ease of use. The improved methods include: (i) *in vitro* synthesis of capped RNA; (ii) *in vitro* translation with biotin incorporation; and (iii) chemiluminescent detection of the biotinylated proteins.

11.2 Technical Procedures

General

All kits and reagents were from Boehringer Mannheim (Mannheim, Germany), unless otherwise mentioned. For the synthesis of capped RNAs *in vitro* the different cDNAs were cloned into suitable transcription vectors downstream of a bacteriophage SP6, T7, or T3 RNA polymerase promoter. Capped mRNA was synthesized *in vitro* with SP6, T3, or T7 Cap-Scribe. The Cap-Scribe sets consist of a phage RNA polymerase and a 5× reaction buffer with NTPs and cap-analogue. Biotin *in vitro* translation was performed with the biotin *in vitro* translation kit. This kit contains a mixture of reticulocyte lysate and biotin-lysine-tRNA^{Lys}, provided in ready-to-use aliquots for single reactions. Dog pancreas microsomes were used for the processing of *in vitro* translated proteins. The translation products were separated by SDS-PAGE minigel and electroblotted onto PVDF-P membrane (Millipore, Bedford, MA). The chemiluminescence Western-blotting kit (biotin/streptavidin) from Boehringer Mannheim was used for detection of biotinylated proteins. The kit contains streptavidin conjugated to horseradish peroxidase, blocking reagent (modified casein), luminol/iodophenol substrate, and starter reagent. The electrophoretic mobility shift assay was performed with the Dig Gel Shift Assay kit (Boehringer Mannheim) and CSPD® (Tropix, Bedford, MA). Chemiluminescent and radioactive signals were recorded by exposure to X-ray film (Du Pont, Wilmington, DE, or Kodak, Rochester, NY). Luciferase activity was measured with the luciferase assay reagent (LAR, Promega, Madison, WI) by a luminometer (EG&G Berthold, Wildbad, Germany).

Materials and buffers

In vitro transcription of capped mRNAs:

- SP6, T3, or T7 Cap-Scribe Set
- Plasmid DNA, with cDNA cloned downstream of either an SP6, a T7, or a T3 RNA polymerase promoter, linearized, phenol/chloroform-extracted, and ethanol precipitated
- 0.2 M EDTA, pH 8.0
- 6 M ammonium acetate
- Ethanol at –20 °C
- 70% ethanol at –20 °C
- Redistilled sterile water, RNAse-free

Chemiluminescent detection of biotin-labeled proteins:

- BM Chemiluminescence Western Blotting Kit
- Maleic acid (Serva)
- NaCl
- Tween® 20
- Methanol

Streptavidin-POD: Reconstitute lyophilized streptavidin-POD (50 units) with 0.1 ml of sterile water. Centrifuge the streptavidin-POD solution for 30 sec to sediment any insoluble material. *Detection solution:* Place substrate solution A (Boehringer Mannheim GmbH) and starting solution B in a 25 °C water bath for 15 min. Mix 2 ml of solution A with 20 µl of solution B and incubate for additional 30 min at room temperature. *Maleic acid solution*: 100 mM maleic acid; 150 mM NaCl, adjust to pH 7.5 with concentrated or solid NaOH. *Blocking reagent stock solution:* 10% (w/v) blocking reagent in maleic acid solution. Dissolve by heating to 70 °C and autoclave. *Blocking solution:* blocking stock solution diluted 1:5 in maleic acid solution. *Streptavidin-POD working solution:* dilute blocking stock solution 1:10 in maleic acid solution and add 0.1 units/ml streptavidin-POD concentrate (dilution 1:5000). *Wash solution:* 0.1% Tween® 20 in maleic acid solution.

Protocol 11.1

In vitro transcription of capped mRNAs

1. Mix on ice to a final volume of 20 µl:
   5 × Cap-Scribe buffer, 4 µl
   Linearized DNA, 0.5 µg
   Add sterile, redistilled H_2O, RNAse-free to 18 µl
   SP6, T7, or T3 RNA polymerase, 2 µl
   (20 units/µl)
2. Centrifuge briefly and incubate at 37 °C for 1 hour.
3. Add 2 µl of 0.2 M EDTA, pH 8.0.
4. Remove 1–2 µl of the reaction mixture, and analyze by agarose gel electrophoresis.
5. Precipitate the transcripts by adding 14 µl of 6 M ammonium acetate and 100 µl of chilled ethanol.
6. Leave at room temperature for 30 min.
7. Centrifuge, wash the pellet with 70% chilled ethanol, and centrifuge again.
8. Dry the pellet and dissolve it in 25 µl of sterile water, RNase-free.
9. Aliquot capped RNA and store at –70 °C.

Protocol 11.2

Biotin *in vitro* translation

1. Thaw vials of biotin *in vitro* translation mix in your hands or in a 30 °C water bath and place immediately on ice.
2. Mix on ice:
   Biotin translation mix, 1 vial (30 µl)
   RNA to be assayed, 0.4–1.5 µg (1–4 µl)
   Make up with H_2O to 50 µl.
4. Mix carefully, do not vortex.
5. Centrifuge briefly.
6. Incubate 1 hour at 30 °C. [Start preparing the SDS-PAGE.]
7. Stop the reaction by placing on ice.
8. Analyze an aliquot by SDS-PAGE and/or store the translation reaction at –20 °C.

Protocol 11.3

SDS-PAGE

1. For minigels: Combine on ice 1–2 µl of the translation assays with 12 µl of 1× loading buffer. Adjust amount and volume for larger gels.
2. Boil samples for 10 min.
3. Centrifuge briefly and load samples in the gel slots.
4. Run either prestained or biotin-labeled protein molecular weight size standards next to the samples.
5. Run gel at <13 V/cm. Higher currents may lead to smearing of the bands.

Protocol 11.4

Western blotting

1. Moisten PVDF membrane with methanol for a few seconds, then soak with transfer buffer for at least 5 min. Soak nitrocellulose briefly in water and then for at least 5 min in transfer buffer.
2. Equilibrate the gel in transfer buffer for 5–10 min.
3. Blot according to standard protocols, for example, semidry blotting at 2.5 mA/cm^2 for 40 min onto a PVDF membrane.

Caution: To avoid damage or contamination of the membrane, always wear gloves when handling.

Protocol 11.5

Chemiluminescent detection of biotin-labeled proteins

1. Incubate membrane for 40 min in 100 ml of blocking solution (2%).
2. Incubate for 30 min in 25 ml of diluted streptavidin-POD solution.
3. Prepare 2 ml of the detection solution as described above.
4. Wash the membrane 4 × 10 min with 100 ml of wash solution.
5. Drain excess liquid off the membrane by placing it protein-side up on a sheet of Whatman 3MM (or similar) paper; do not let the membrane dry.
6. Enclose the membrane in a sealable plastic bag cut to the size of the membrane.
7. Add 1 ml of the chemiluminescence substrate mix and heat-seal the bag, avoiding air bubbles.
8. Distribute the substrate evenly in the bag, expose to an X-ray film for a few seconds or a few minutes.

Incubate the individual blots in separate dishes, and place them protein-side downward.
The volumes of the solutions are calculated for a membrane size of 100 cm^2 and should be adjusted to accommodate other membrane sizes.

11.3 Results and Discussion

The principle of biotin labeling of nascent polypeptides during *in vitro* translation is schematically shown in Figure 11.1: lysine is coupled to $tRNA^{Lys}$ from brewer's yeast using rat liver aminoacyl-tRNA synthetases. In our hands, yeast tRNA was aminoacylated with lysine to a higher extent than tRNA from *E. coli* (data not shown). The lysine is modified at its ε-NH_2-group with D-biotinoyl-ε-aminocaproic acid-*N*-hydroxysuccinimide ester. The biotin-lysine-$tRNA^{Lys}$ can be added directly to a translation system. In the biotin *in vitro* translation kit, the biotin-lysine-$tRNA^{Lys}$ and the rabbit reticulocyte extract are provided premixed and in ready-to-use aliquots of 30 µl. The translation mixture also contains tRNAs, "energy mix," consisting of creatine phosphate, creatine phosphokinase, GTP and ATP, dithiothreitol, and all the other 19 amino acids except lysine. In addition, the required minimum

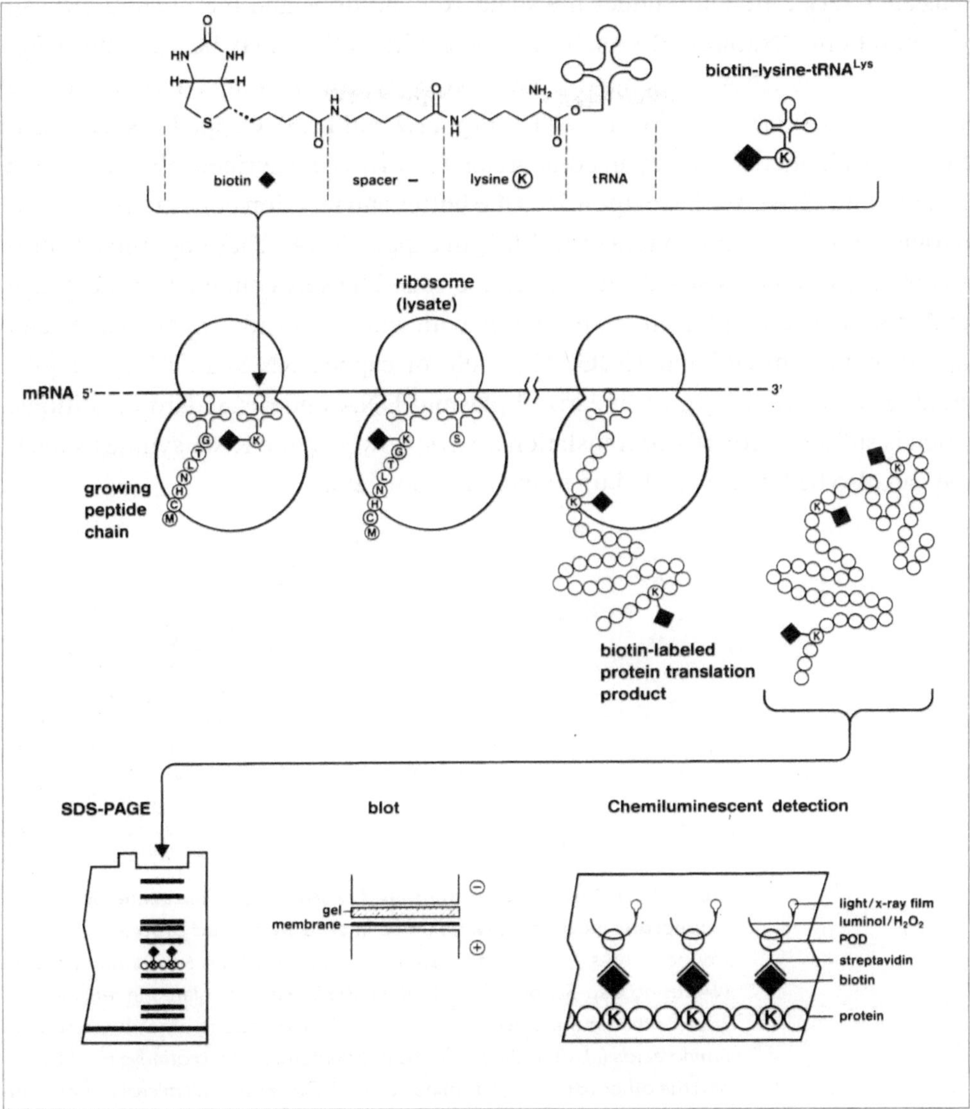

Figure 11.1 Principle of biotin in vitro translation and chemiluminescent detection

concentration of potassium acetate and magnesium acetate is included in the mix. The assay is started by addition of an RNA (0.5–2 μg) with a translatable open reading frame. In principle, any type of RNA can be translated into its coded proteins. Although the yield of biotin-labeled translation product depends on a variety of factors, for example, concentration of the RNA and salt conditions, most RNAs are translated very efficiently under the standard conditions and the preadjusted salt concentrations. "Natural" RNAs, for example, viral RNAs, may be translated (Fig. 11.2) as well as RNAs generated from cDNAs by *in vitro* transcription with the phage RNA polymerases SP6, T3, or T7 (Fig. 11.2 and 11.3). Capped RNA is translated with a higher efficiency in comparison to *in vitro* transcribed RNA without a cap structure (2, 4). We have optimized the buffer and reaction conditions of *in vitro* transcription for a high yield of full-length capped RNA. These optimized components are offered as so-called Cap-Scribe sets. The sets contain 1 of the phage RNA polymerases and a premixed buffer with ribonucleoside triphosphates and cap-nucleotide, $m^7G(5')ppp(5')G$. The yield of capped RNA is ~10 µg of full-length RNA from 0.5 µg of linearized plasmid DNA under standard conditions, which is sufficient for 10–20 translation assays; however, the RNA synthesis reaction can also be linearly scaled up to increase the yield.

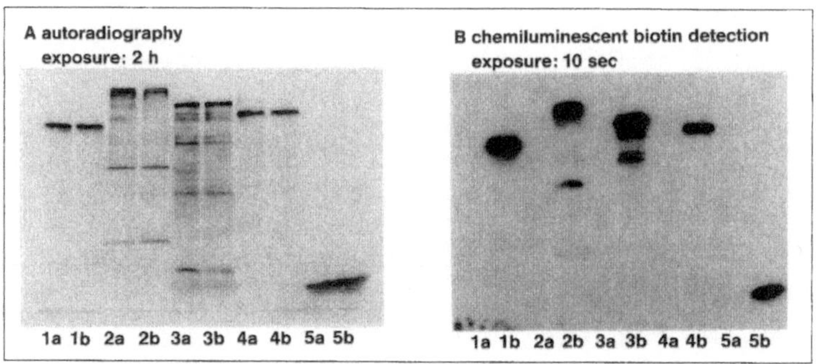

Figure 11.2 Biotin in vitro translated proteins have the same electrophoretic mobility as radioactive in vitro translated proteins

Capped RNAs 1, 3, 4, and 5 were generated by in vitro transcription of cDNAs using Cap-Scribe. Aliquots of the RNAs were translated in reticulocyte lysate either in the presence of [^{35}S]methionine and the other 19 unlabeled amino acids (a), or in the presence of [^{35}S]methionine, biotin-lysine-tRNA^{Lys}, and the other 18 unlabeled amino acids (b). One and a half microliters of each of the translation reactions was run on a 12% SDS-polyacrylamide gel. After semidry blotting to a PVDF membrane, the radioactive labeled translation products (a and b) were detected by autoradiography (A). The biotin-labeled translation products (b) were then detected by chemiluminescence using streptavidin-peroxidase and luminol/iodophenol as substrate (B). a: in vitro translations in reticulocyte lysate with [^{35}S]methionine plus the other 19 amino acids; b: in vitro translations in reticulocyte lysate with [^{35}S]methionine and biotin-lysine-tRNA^{Lys} plus the other 18 amino acids; 1: γ-globulin RNA; 2: Brome Mosaic Virus RNA; 3: luciferase RNA; 4: tissue plasminogen activator RNA; 5: β-globin RNA.

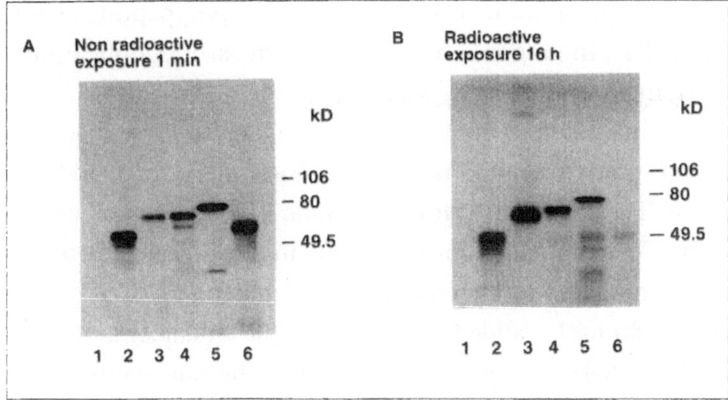

Figure 11.3 Comparison of biotin* in vitro *translation with chemiluminescent detection and radioactive* in vitro *translation with [^{35}S]methionine

Capped RNA transcripts were used for nonradioactive in vitro translation without further purification. One microliter of each of the Cap-Scribe products was added to 1 vial (30 µl) of the biotin in vitro translation mix. To the control assay RNA was not added. After incubation for 1 hour at 30 °C, the reaction was stopped on ice. 2 µl of the translation assays was run on an SDS polyacrylamide gel and electroblotted onto a PVDF-P membrane. The biotinylated proteins were detected by a streptavidin horseradish peroxidase conjugate and visualized by chemiluminescence with luminol/iodophenol. The luminescent signals were recorded by a 1-min exposure to X-ray film (A). For comparison, 2 µl of the same Cap-Scribe RNAs was translated in a reticulocyte lysate in the presence of 20 µCi of [^{35}S]methionine. Two microliters of the translation assays was separated by SDS-PAGE. The gel was exposed to X-ray film for 16 hours (B). 1: control without RNA; 2: γglobulin; 3: tissue plasminogen activator, tPA; 4: transcription factor CTF1; 5: firefly luciferase; 6: factor IX.

Upon addition of an RNA with a translatable open reading frame to a biotin *in vitro* translation system, the encoded protein is synthesized with the incorporation of the biotin-labeled lysine, provided by the biotin-lysine-tRNA^{Lys}, at some of the lysine residues. The electrophoretic mobility of biotin *in vitro* translated proteins does not significantly differ from the electrophoretic mobility of the same proteins translated in the presence of [^{35}S]methionine (Fig. 11.2). Thus, the number of biotin-labeled lysine residues incorporated per protein molecule must be relatively low. The biotin *in vitro* translation mixture contains some endogenous (unlabeled) lysine as well as tRNA^{Lys}. It is probable that a single protein may carry only 1 or a few biotin residues and only every third or fourth lysine in the polypeptide chain may be labeled with biotin. This finding is in accordance with earlier publications on fluorescence-labeled lysine-$tRNA^{Lys}$ (5).

After the *in vitro* translation reaction, the translation products are separated on an SDS-polyacrylamide gel, transferred to a membrane, and finally detected by a chemiluminescent reaction. The "Western" blot membrane is first blocked with a solution of modified casein. Then blotted biotinylated proteins are detected by binding of streptavidin that is conjugated to the enzyme horseradish peroxidase. The chemiluminescent substrate luminol is added together with the enhancer iodophenol and hydrogen peroxide as starter reagent. The enzymatic luminol-peroxidase reaction produces a strong chemiluminescent signal, which is recorded by a short exposure (few seconds–10 min) to X-ray film. The luminescent signal lasts for ~60–90 min at room temperature, allowing multiple exposures to be made during this period. Using reticulocyte lysate, biotin *in vitro* translation produces equivalent results in comparison with radioactive *in vitro* translations with [^{35}S]methionine if the amount of mRNA used for translation in both systems and the aliquots loaded on the gels are the same (Fig. 11.3), suggesting that this nonradioactive method is a perfect substitute for radioactive *in vitro* translation. In addition, the biotin *in vitro* translation method is faster than radioactive methods (Table 11.1) and avoids the inconvenience, hazards, and stability problems that are associated with radioactivity.

It has previously been demonstrated that proteins labeled chemically with biotin-*N*-hydroxysuccinimide ester under mild conditions are still able to bind to cell surface receptors (26). We have performed several experiments to investigate whether biotin labeling by *in vitro* translation interferes with the biological activ-

Table 11.1. Flow diagram

Step	Duration (min)
Translation	60
SDS-PAGE	100
Blotting	<60
Blocking	40
Streptavidin-POD Binding	30
Washing	40
Substrate incubation/exposure	10
Total	340

ity of the respective translation products. In the first experiment we studied whether *in vitro* translated biotinylated secretory precursor proteins are correctly processed by dog pancreas microsomes (1, 21, 28). The signal peptides of bovine preprolactin ([ppl], 18) and a synthetic presecretory protein (ppcecDHFR), a fusion hybrid of silkworm preprocecropin A and mouse dihydrofolate reductase (24, 25), synthesized *in vitro* with biotin-lysine-$tRNA^{Lys}$, were processed by the signal peptidase of dog pancreas microsomes, giving rise to proteins of lower molecular mass (Fig. 11.4). As expected, yeast prepro-α-factor (pp*αf*) was glycosylated by microsomes to a higher molecular mass product. The so-called sequestration analysis (3), that is, treatment of the translation reactions with proteinase K in the presence or absence of the detergent Triton X-100, confirmed the translocation of the presecretory proteins into the microsomes. Thus, incorporation of biotinylated lysine residues does not interfere with transport of various precursor proteins into dog pancreas microsomes. Typical covalent modifications are observed which occur in the microsomal lumen concomitant with transport, such as cleavage of amino terminal signal sequences by signal peptidase, and core glycosylation by oligosaccharyl transferase.

Another experiment was performed to clarify whether biotin-labeled *in vitro* translated transcription factors are capable of binding to their target DNAs. cDNAs of the transcription factors Oct1 and Oct4 were transcribed into capped RNAs and subsequently translated with the biotin *in vitro* translation kit or into unmodified proteins. The specific binding of these proteins to their target DNA recognition sequence was assayed by a nonradioactive electrophoretic mobility shift assay (EMSA) with a digoxigenin-labeled oligonucleotide (Fig. 11.5). This experiment demonstrated that biotin-labeled translation products are able to recognize and bind to their target DNA recognition sequence. On the blot, signals from unlabeled translation products were stronger than from the biotin-labeled translation products, probably due to a higher yield of protein without biotin labeling.

Finally, we asked whether incorporation of biotinylated lysine residues interferes with folding of a model protein into its native, enzymatically active state. Firefly luciferase was synthesized with the biotin *in vitro* translation kit as well as in a standard translation assay without biotinylated lysine. After various incubation times aliquots were withdrawn from the translation reactions, and the enzymatic activity of luciferase was measured. As shown in Figure 11.6, the kinetics of formation and yield of luciferase were more or less identical under the 2 conditions.

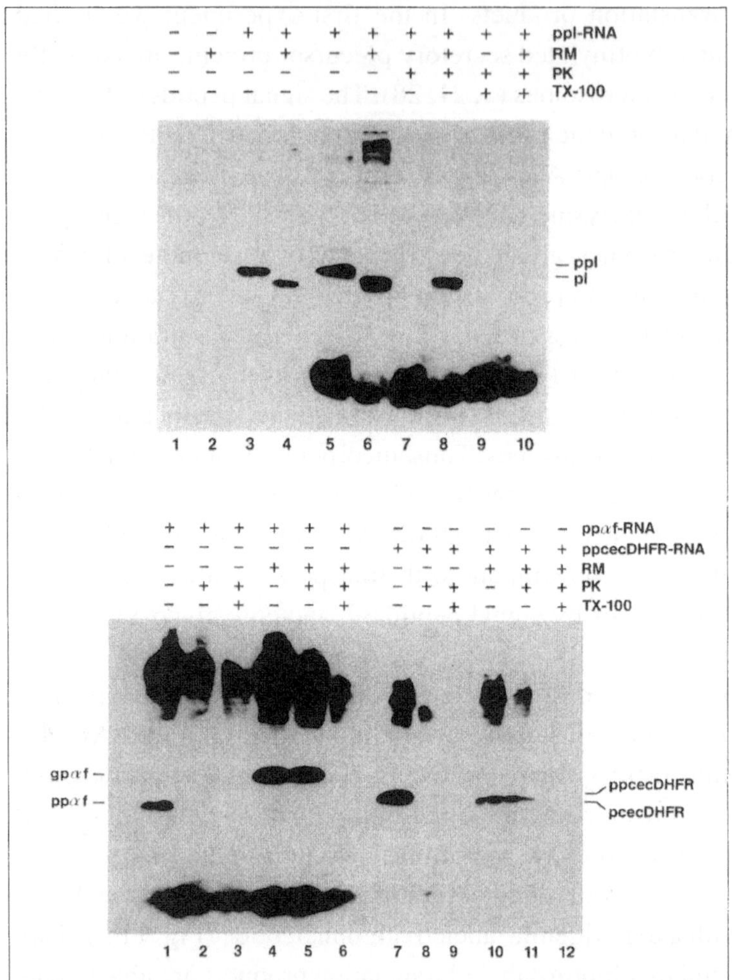

Figure 11.4 Transport and processing of biotin in vitro translated preprolactin, prepro-α-factor, and preprocecropin A-dihydrofolate reductase hybrid protein with dog pancreas microsomes

A: Lanes 1–4: Thirty microliters of biotin in vitro translation mix was divided into 4 aliquots of 6 µl each. To aliquots 3 and 4, 3.6 or 2.8 µl of water and 0.4 µl of in vitro transcribed preprolactin-RNA was added, whereas aliquots 1 and 2 received only 4 or 3.2 µl of water, respectively. Microsomes (0.8 µl) were added to aliquots 2 and 4. The assays were incubated for 1 hour at 30 °C. The reaction was terminated by addition of sample buffer. Half of the final volume was analyzed by gel electrophoresis, Western blotting, and chemiluminescent detection as described in the protocol. Exposure time was 3 min. Lanes 5–10: Thirty microliters of biotin in vitro translation mix was supplemented with 2 µl of in vitro transcribed preprolactin-RNA plus 14 µl of water and divided in 2 aliquots of 23 µl each. Then 2 µl of water (lanes 5, 7, 9) or 2 µl of microsomes (lanes 6, 8, 10) was added, and the assays were incubated for 1 hour at 30 °C. The reaction was terminated on ice. Each aliquot was further divided into 3 aliquots of 7.5 µl and supplemented with 5 µl of 0.25 M sucrose solution plus either 1 µl of proteinase K solution (1 mg/ml) (lanes 7 and 8) or proteinase K solution plus 1.5 µl of Triton X-100 (5%, v/v) (lanes 9 and 10). The different aliquots were made up to 15 µl with water and incubated for 1 hour on ice. The reactions were terminated by the addition of 2 µl of phenylmethylsulphonyl fluoride solution (0.1 M in ethanol) and were incubated for 5 min on ice. Then sample buffer was added, and half of the final volume was analyzed by gel electrophoresis, Western blotting, and chemiluminescent detection as described in the procotol. Exposure time was 8 min.

B: Lanes 1–6: Thirty microliters of biotin in vitro translation mix was supplemented with 4 µl of in vitro transcribed prepro-α-factor RNA plus 14 µl of water and divided in 2 aliquots of 24 µl each. One microliter of water (lanes 1–3) or 1 µl of microsomes (lanes 4–6) was added. The reactions were incubated for 1 hour at 30 °C and then stopped on ice. Further processing of the samples with proteinase K (lanes 2 and 5) or proteinase K plus Triton (lanes 3 and 6) and analysis of the different aliquots was as described above. Exposure time was 4 min. Lanes 7–12: Thirty microliters of biotin in vitro translation mix was supplemented with 4 µl of in vitro transcribed preprocecropin A-dihydrofolate reductase RNA plus 12 µl of water and divided in 2 aliquots of 23 µl each. Then 2 µl of water (lanes 7, 8, 9) or 2 µl of microsomes (lanes 10, 11, 12) was added, and the assays were incubated for 1 hour at 30 °C. The reaction was terminated on ice. Further processing of the samples with proteinase K (lanes 8 and 11) or proteinase K plus Triton (lanes 9 and 12) and analysis of the different aliquots was as described above. Exposure time was 1 min. ppl: preprolactin; pl: prolactin; ppαf: prepro-α-factor; gpaf: glycosylated prepro-α-factor; pcecDHFR: procecropin A-dihydrofolate reductase hybrid protein; ppcecDHFR: preprocecropin A-dihydrofolate reductase hybrid protein; RM: dog pancreas microsomes; PK: proteinase K; TX-100: Triton X-100.

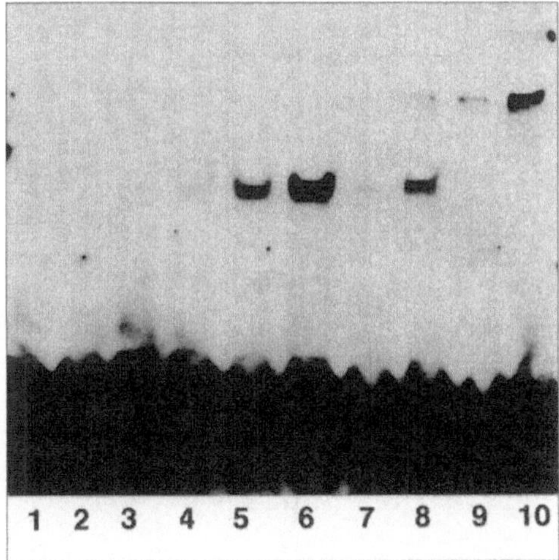

Figure 11.5 Electrophoretic mobility shift assay with digoxigenin-labeled oligonucleotides and biotin-labeled *in vitro* translated transcription factors

cDNAs of transcription factors Oct1 and Oct4 were transcribed into capped RNA using Cap-Scribe. Two microliters of each was translated with the biotin in vitro translation kit. Parallel in vitro trans-

To Figure 11.5 (continued)

lations were performed in rabbit reticulocyte lysate without biotin labeling in the presence of 20 unlabeled amino acids. The ability of the biotin-labeled as well as the unlabeled translation products to bind to their DNA recognition sequences was determined by incubation of 5-µl aliquots of the translation assays with a digoxigenin-labeled double-stranded oligonucleotide containing the (identical) recognition sequence of Oct1 and Oct4. Specific interaction of transcription factors with the oligonucleotide was assayed by an electrophoretic mobility shift assay. The oligonucleotide-protein complexes were detected after blotting to a positively charged nylon membrane by antidigoxigenin alkaline phosphatase and CSPD®. The chemiluminescent signal was recorded by a 20-min exposure to X-ray film. 1: Control biotin translation without RNA; 2: control unlabeled translation without RNA; 3: biotin-labeled Oct4, competition with 400 ng of unlabeled oligonucleotide; 4: unlabeled Oct4, competition with 400 ng of unlabeled oligonucleotide; 5: biotin-labeled Oct4; 6: unlabeled Oct4; 7: biotin-labeled Oct4 + biotin-labeled Oct1; 8: unlabeled Oct4 + unlabeled Oct1; 9: biotin-labeled Oct1; 10: unlabeled Oct1.

Biotin *in vitro* translation is a nonradioactive alternative with many advantages over radioactive *in vitro* translation assays: the method avoids the use of unstable and hazardous radioactive tracers, yields equivalent results regarding sensitivity and resolution, and is much faster than the radioactive procedure. This nonradioactive method can replace radioactive *in vitro* translations for all applications tested up to now, including assays where the biological function of the *in vitro* translated proteins is of relevance.

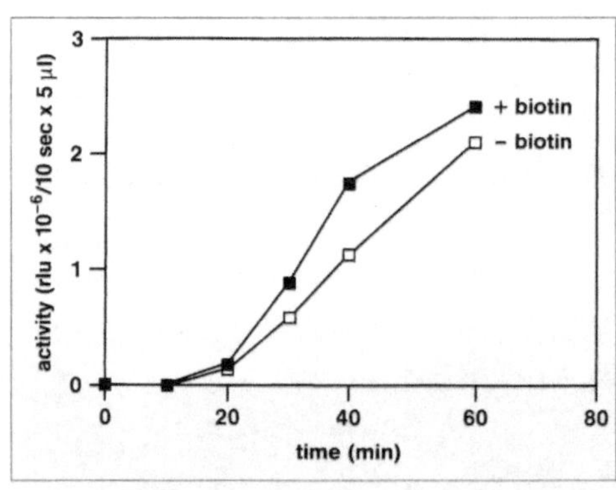

Figure 11.6 Biotin** in vitro **translated firefly luciferase folds into an active enzyme

Four microliters of in vitro transcribed firefly luciferase RNA were in vitro translated in reticulocyte lysate in the presence of biotin-lysine-tRNA^{Lys} or with 20 unlabeled amino acids at 30 °C. At the indicated time 5 µl was withdrawn, and the luciferase activity was measured after addition of 50 µl of luciferase assay reagent in a luminometer as relative light units.

Biotin-labeled translation products can also be immobilized to solid supports, for example, to magnetic beads covered with streptavidin, avidin, or antibiotin antibodies. This feature enables purification and further study of the biotin-labeled polypeptides.

11.4 Troubleshooting

The translation products are separated on an SDS-polyacrylamide gel, transferred to a membrane, and finally detected by chemiluminescent reaction. Although we prefer SDS-PAGE minigels and electroblotting to a PVDF membrane, any type of polyacrylamide gel electrophoresis (denaturing or nondenaturing, 1-D, 2-D), blotting method, and membrane (PVDF, nitrocellulose, or nylon) can be used. The blotted biotinylated proteins are detected by binding of streptavidin that is conjugated to the enzyme horseradish peroxidase. Alternatively, one can also use streptavidin conjugated to the enzyme alkaline phosphatase in combination with appropriate color or chemiluminescent substrates. The color detection with 5-bromo-4-chloro-3-indolyl phosphate (BCIP, X-phosphate) and nitroblue tetrazolium chloride (NBT) is not as sensitive and rapid as the peroxidase/luminol chemiluminescent system, but can also be applied on nitrocellulose and PVDF membranes. However, with the chemiluminescent alkaline phosphatase substrates CSPD® (Tropix, Bedford, MA), LumigenPPD™, or Lumi-Phos™ (Lumigen, Detroit, MI) nylon membranes have to be used for blotting to achieve high sensitivity. Nitrocellulose or PVDF membranes can be used for the alkaline phosphatase-catalyzed chemiluminescent detection only after treatment with macromolecular hydrophobic blocking agents like Nitro-Block™ (Tropix). Since PVDF membranes are best for protein blotting, the peroxidase/luminol chemiluminescent detection method produces relatively strong signals with very low background in the shortest possible time. We therefore prefer this system for detection.

The standard radioactive translation kit and the Biotin Translation Kit appear to be equally effective in protein synthesis. In the examples shown in Figure 11.4, the background which was observed by chemiluminescent detection was higher than the background typically observed after incorporation of radioactive amino acids. However, we note that in the case of the Biotin Translation Kit the signal-to-background ratio can be improved by applying larger aliquots of the translation reactions onto the SDS-PAGE.

Acknowledgments

We thank T.V. Kurzchalia for helpful support and advice, W. Ankenbauer for the DIG gel shift assay procedure, H. Schoeler for the Oct1 and Oct4 cDNA clones, and B. Patel for critical reading of the manuscript.

References

1 Blobel, G. and Dobberstein, B. (1975) *J. Cell. Biol.* **67**, 852–862.

2 Clemens, M.J. (1984) Translation of Eukaryotic Messenger RNA in Cell-Free Extracts. *In*: B.D. Hames and S.J. Higgins (eds) *Transcription and Translation: A Practical Approach*, IRL Press, Oxford, Washington, DC, pp 231–270.

3 Connolly, T., Collins, P. and Gilmore, R. (1989) *J. Cell. Biol.* **108**, 299–307.

4 Craig, D., Howell, M.T., Gibbs, C.L., Hunt, T. and Jackson, R.J. (1992) *Nucleic Acids Res.* **20**, 4987–4995.

5 Crowley, S.C., Reinhart, G.D. and Johnson, A.E. (1993) *Cell* **73**, 1101–1115.

6 DeVries, J.K. and Zubay, G. (1969) *J. Bacteriol.* **97**, 1419–1425.

7 Erickson, A.H. and Blobel, G. (1983) Cell-Free Translation of Messenger RNA in a Wheat Germ System. *In*: S. Fleischer and B. Fleischer (eds) *Methods in Enzymology*, vol. 96, Academic Press, New York, pp 38–50.

8 Görlich, D., Prehn, S., Hartmann, E., Herz, J., Otto, A., Kraft, R., Wiedmann, M., Knespel, S., Dobberstein, B. and Rapoport, T.A. (1990). *J. Cell. Biol.* **111**, 2283–2294.

9 Görlich, D., Kurzchalia, T.V., Wiedmann, M. and Rapoport, T.A. (1991) Probing the Molecular Environment of Translocating Polypeptide Chains by Crosslinking. *In*: A.M. Tartakoff (ed.) *Laboratory Methods in Vesicular and Vectorial Transport*, Academic Press, New York, pp 37–58.

10 Görlich, D., Prehn, S., Hartmann, E., Kallies, K.-U. and Rapoport, T.A. (1992) *Cell* **71**, 489–503.

11 High, S., Görlich, D., Wiedmann, M., Rapoport, T.A. and Doberstein, B. (1991) *J. Cell. Biol.* **113**, 35–44.

12 Jackson, R.J., Campbell, E.A., Herbert, P. and Hunt, T. (1983) *Eur. J. Biochem.* **131**, 289–301.

13 Johnson, A.E., Woodward, W.R., Herbert, E. and Menninger, J.R. (1976) *Biochemistry* **15**, 569–575.

14 Johnson, A.E., Miller, D.L. and Cantor, C.R. (1978) *Proc. Natl. Acad. Sci. USA* **75**, 3075–3079.

15 Johnson, A.E. and Slobin, L.I. (1980) *Nucleic Acids Res.* **8**, 4185–4200.

16 Krieg, P. and Melton, D. (1984) *Nucleic Acids Res.* **12**, 7057–7070.

17 Krieg, U.C., Walter, P. and Johnson, A.E. (1986) *Proc. Natl. Acad. Sci. USA* **83**, 8604–8608.

18 Kurzchalia, T.V., Wiedmann, A., Girshovich, A.S., Bochkareva, E.S., Bielka, H. and Rapoport, T.A. (1986) *Nature* **320**, 634–636.

19 Kurzchalia, T.V., Wiedemann, M., Breter, H., Zimmermann, W., Bauschke, E. and Rapoport, T.A. (1988) *Eur. J. Biochem.* **172**, 663–668.

20 Pelham, H.R.B. and Jackson, R.J. (1976) *Eur. J. Biochem.* **67**, 247–256.

21 Perara, E., Rothman, R.E. and Lingappa, V.R. (1986) *Science* **232**, 348–352.

22 Pratt, J.M. (1984). Coupled Transcription-Translation in Procaryotic Cell-Free Systems. *In*: B.D. Hames and S.J. Higgins (eds) *Transcription and Translation: A Practical Approach*, IRL Press, Oxford, Washington, DC, pp 179–209.

23 Roberts, B.E. and Paterson, B.M. (1973) *Proc. Natl. Acad. Sci. USA* **70**, 2330–2334.

24 Schlenstedt, G., Gudmundsson, G.H., Bomann, H.G. and Zimmermann, R. (1990) *J. Biol. Chem.* **265**, 13960–13968.

25 Schlenstedt, G., Gudmundsson, G.H., Bomann, H.G. and Zimmermann, R. (1992) *J. Biol. Chem.* **267**, 24328–24332.

26 Sivaram, P., Wadhwani, S., Klein, M.G., Sasaki, A. and Goldberg, I.J. (1993) *Anal. Biochem.* **214**, 511–516.

27 Tuite, M.F., Plesset, J., Moldave, K. and McLaughlin, C.S. (1980) *J. Biol. Chem.* **255**, 8761–8766.

28 Walter, P. and Blobel, G. (1983) Preparation of Microsomal Membranes for Cotranslational Protein Translocation. *In*: S. Fleischer and B. Fleischer (eds) *Methods in Enzymology*, Vol. 96, Academic Press, New York, pp 557–561.

29 Wiedmann, M., Kurzchalia, T.V., Bielka, H. and Rapoport, T.A. (1987) J. *Cell. Biol.* **104**, 201–208.

30 Zubay, G. (1973) *Annu. Rev. Genet.* **7**, 267–287.

12

Nonradioactive Detection of Nucleic Acids with Biotinylated Probes

Rainer Löw and Thomas Rausch

Summary

Methods for nonradioactive detection of nucleic acids have replaced ^{32}P-based techniques in many laboratories. Improvements in sensitivity allow the detection of single-copy genes in Southern blots and low abundance mRNAs in Northern blots. Labelling of (c)DNA probes involves replacement of dTTP by either digoxigenin-dUTP or biotinylated dUTP. Here we describe experimental protocols for Northern blot analysis and for genomic Southern blot using a rapid alkaline transfer technique, followed by procedures for hybridization and blot development with biotinylated probes. A protocol for synthesis of biotin-labeled probes by polymerase chain reaction (PCR) is included. The technical difficulties encountered with densitometric signal quantitation of nonradioactive blots are discussed, and procedures for relative and absolute quantitation of bound probe are presented.

12.1 Introduction

Technical improvements in methods used for nonradioactive detection of nucleic acids have convinced an increasing number of laboratories to replace traditional ^{32}P-based detection procedures by nonradioactive detection with either digoxigenin- (1) or biotin-labeled probes (2). The technical advances include, in particular, (i) the availability of a number of new substrates for alkaline phosphatase that provide high sensitivity for signal detection (chemiluminescence substrates, e.g., CSPD, CPD, or CDP-Star from Tropix or equivalent reagents from other suppliers), (ii) the development of reliable experimental protocols for detection of single-copy genes in complex eukaryotic genomes (3), and (iii) the extension of the nonradioactive procedure to the Northern blot technique (4). If quantitative information on signal intensity is required, nonradioactive detection procedures do pose some difficulties not encountered with the traditional ^{32}P-based technique, in which liquid scintillation counting of excised bands allows rapid quantitative comparisons. However, when proper controls are included (see below), the densitometric quantitation of signals obtained with biotinylated probes is possible. When the probe is synthesized by PCR with a biotinylated primer instead of random incorporation of bio-16-dUTP, even the absolute number of bound probe molecules can be estimated. We present a brief survey of techniques used for nucleic acid detection with biotinylated probes. Particular emphasis is paid to the Northern blot technique and the possibility of obtaining quantitative information from nonradioactive blots. The protocols presented will allow the user to introduce the nonradioactive method for detection of DNA and RNA in most practical applications.

12.2 Technical procedures

PCR-labeling of (c)DNA probes

Probe synthesis by PCR yields a product of defined size with dTTP being randomly replaced by bio-16-dUTP (5, 6). Label incorporation is dependent on (i) the ratio of bio-16-dUTP/dTTP in the reaction mixture, and (ii) the discrimination between bio-16-dUTP and dTTP by the particular type of Taq polymerase used. The yield and quality (label incorporation) of PCR-labeled probes can be assessed by com-

parison with the corresponding unlabeled PCR product (see below).
The primers used for probe synthesis by PCR may be directed either against the sequence of interest itself or, alternatively, when a clone is available, against the insert-bordering vector sequences. The latter has the advantage that the same set of primers can be used for different inserts cloned into the same vector. However, the use of vector-based primers excludes the use of the probe for colony hybridizations and related techniques due to additional vector-vector hybridization.

- A typical reaction includes in a total volume of 100 µl:
- 1 pg–1 ng template
- 25–50 pmol of each primer
- 1× reaction buffer (dilution of specific DNA polymerase buffer, usually supplied as a 10× stock)
- 25 µM each dNTP (10 µl of a 1 mM nucleotide mix: 250 µM each dATP, dCTP, dGTP, and 125 µM each of bio-16-dUTP and dTTP)
- 1.0–2.5 units of thermostabile DNA polymerase (depending on the supplier; we use 2.5 units of Amersham thermostabile DNA polymerase)

Reaction conditions

In principle, every primer/template combination has its specific optimum annealing temperature. The standard program used in our lab for inserts (300–700 bp) cloned into pBluescript SK+ and T3 and T7 as sense and antisense primer, respectively (Stratagene), is the following: Initial denaturation at 94 °C for 5 min, followed by 30 cycles of 94 °C/30 sec-55 °C/60 sec–72 °C/30 sec, with the final 72 °C step extended for 5 min. Under these conditions a PCR reaction yields about 2 µg of labeled probe. Probe purification, that is, removal of Taq polymerase, primers, and nonincorporated bio-16-dUTP is not necessary, except in those cases where additional nonspecific amplification products are formed. Under these circumstances the probe can be purified according to standard procedures (7, 8).

As an obligate quality control, a parallel PCR reaction should be run with bio-16-dUTP replaced by dTTP. Aliquots of the reaction products should be analyzed on a 1–2% TBE agarose gel or a 10% acrylamide gel. Usually the apparent size of the probe should exceed the product of the control reaction by about 10–20%, due to the incorporation of bio-16-dUTP. The biotinylated probe runs as a diffuse band due to the random incorporation of label. The above reaction conditions are optimized for the Amersham enzyme and bio-16-dUTP as provided by Boehringer Mannheim. For more information the reader is referred to the Troubleshooting section. The quality of the biotinylated probe should be further checked by a dot blot against the unlabeled PCR product. The detection limit for probes of 300–700 bp (optimum probe size) should be between 1 pg and 0.1 pg (note that probes below 200 bp may not be able to detect target below 1 pg).

PCR probes are stable for up to 1 year at –20 °C or 6 months at 4 °C. Hybridization solutions with biotinylated probes (see below) can be used several times. They should be stored at 4 °C, where they are stable for at least 6 months.

Northern blot

Until recently, nonradioactive Northern blots with total RNA have been hampered by limited sensitivity and high nonspecific background labeling. These problems could be reduced by using mRNA instead of total RNA; however, mRNA purification involves considerable extra cost. The protocol we have recently developed (4), and which we describe below, has allowed us to detect consistantly low abundance mRNAs (<0.01% of total mRNA) in total RNA samples.

As mRNAs may have considerable secondary structure involving strong intramolecular hybridization, a critical step for high sensitivity seems to be complete denaturation during capillary transfer and cross-linking to the membrane. To achieve this we use a rapid capillary transfer step (buffer-soaked sponge as buffer reservoir) and mildly alkaline conditions (10 mM NaOH, 5× SSC, pH 11.6).

Gel electrophoresis, blotting, and fixation

Total RNA is extracted according to standard procedures (9) and separated on a 1.2–1.6% formaldehyde-agarose gel (1× MOPS, 2.2 M formaldehyde; for standard buffer recipes, e.g., MOPS, SSC, SSPE, see refs. 7 and 8) at 5.0–7.5 V/cm. After electrophoresis the gel is incubated twice for 10 min in 5× SSC, 10 mM NaOH. For the capillary transfer a sponge serves as buffer reservoir. The sponge assures rapid mass flow through the entire gel area. The weight on the top of the "blotting sandwich" should be 2–3 g/cm^2. Transfer should be complete after 60 min (>90% of RNA removed from the gel). After transfer the membrane is briefly rinsed (2 min) in 5× SSC, air-dried for 2 min at room temperature and UV cross-linked at 1200 μJ/cm^2 (conditions for cross-linking may vary for different types of membranes; check supplier for appropriate conditions). After cross-linking, the membrane should be dried completely (minimum 1–2 hours) at room temperature before starting the hybridization.

Membranes

Our procedure works well with any of the following membranes: Flash membrane (neutral nylon; Stratagene), Hybond (neutral and positive nylon, Amersham); Tropilon (positive nylon, Tropix). However, neutral nylon membranes give the best signal-to-noise ratio.

Hybridization

Hybridization solutions may be used several times (storage at 4 °C), but care has to be taken to avoid loss of water during repeated denaturation. Prehybridization is performed in probe-free hybridization solution at 37–42 °C for 30–60 min, followed by overnight hybridization at the same temperature (for more details see ref. 10 and literature cited therein).

Hybridization solution

- 30–50% formamide
- 1–5% SDS
- 6% polyethylene glycol 6000
- 1 M NaCl
- 250 μg/ml sheared and denatured salmon sperm DNA
- 0.1 nM probe (denatured: 5 min, 95 °C)

After the membrane is removed from the hybridization solution, it is washed twice for 10 min in 2× SSPE, 0.5% SDS at room temperature (low stringency wash), followed by incubation in 0.2× SSPE, 0.5% SDS at 45–65 °C (depending on the stringency) for 30 min.

Blot development

The following steps require the SDS-resistant (strept)avidin-alkaline phosphatase conjugate (AP conjugate), which at present is available only from Tropix. If other AP conjugates are used, SDS should be replaced by 0.2–0.3% Tween 20.
Following the high-stringency wash, the membrane is incubated twice for 5 min and once for 30 min in blocking buffer. Conjugation is performed in blocking buffer supplemented with AP conjugate (1:6000) for 60 min. After conjugation the blot is washed once in blocking buffer, 4 times in wash buffer, followed by 2 washes in assay buffer. All wash steps are done for 3–5 min. Finally, the membrane is incubated in substrate buffer (assay buffer plus substrate, e.g., CSPD) for 5–15 min. Substrate dilution should follow the instructions of the supplier (note, however, that CPD-Star may be used at 10-fold dilution of the supplier's "ready-to-use" solution; this yields a signal strength comparable to that of CSPD while reducing the cost of substrate considerably).

For detection of chemiluminescence, X-ray films (e.g., Kodak X-Omat or Hyperfilm, Amersham) are usually exposed for 30–60 min, but exposure can be extended up to several hours without problems due to nonspecific background staining. Even an overnight exposure at 4 °C is possible, and yields signals comparable to a 2–3 hour exposure at room temperature.

All steps are performed at room temperature (23–25 °C) in clean plastic containers of appropriate sizes. For a 100-cm^2 blot, usually 200 ml of each solution is sufficient. The conjugation step needs 10–15 ml of solution. For this step a separate container should be used. In order to save expensive AP conjugate and substrate, we have introduced 2 modifications. First, blots can be conjugated simultaneously if care is taken to soak each individual blot in AP-conjugate solution before placing the next blot on top. Second, for substrate incubation we routinely use plastic bags cut open at the sides. After this the top is folded back and the blot placed upside up with 1.5 ml (for a 100-cm^2 membrane) pipetted on its surface. The top of the bag is then used to cover the membrane, thereby evenly spreading the substrate solution, and the blot is ready for incubation.

Solutions

Blocking buffer: 1× PBS, 0.5% SDS, 0.2% casein (in the Tropix "Southern Light" kit the casein is called "I-block." If you use a different source of casein, make sure it is devoid of phosphatase contamination).

Conjugation buffer: 1× PBS, 0.5% SDS, 0.2% casein, 1:6000 dilution of AP conjugate (AP conjugate is included in the Tropix "Southern Light" kit. If you use a different source of AP conjugate, check for appropriate dilution and buffer condition of the conjugate).
Wash buffer: 1× PBS, 0.5% SDS.
Assay buffer: 0.1 M diethanolamine, 1 mM $MgCl_2$, pH 10.0.
Substrate buffer: 0.1 M diethanolamine, 1 mM $MgCl_2$, pH 10.0, 1:100 dilution of CSPD concentrate (Tropix, 25 mM).

Protocol 12.1 **Blot development**

1. 2× wash in 2× SSPE, 0.5% SDS, 15 min at room temperature (low-stringency wash)
2. 1× wash in 0.2× SSPE, 0.5% SDS, 30–60 min at appropriate temperature (high-stringency wash)
3. 2× wash in blocking buffer, 5 min
4. 1× wash in blocking buffer, 30 min
5. Incubation in conjugation buffer, 60 min (use separate container)
6. 1× wash in blocking buffer, 5 min
7. 4× wash buffer, 5 min
8. 2× wash in assay buffer, 5 min
9. Incubation in substrate buffer, 5–15 min

Blot reprobing

Membranes can be reprobed 3–4 times before the signal/background ratio will become unacceptable. For reprobing, the membrane is washed once in 0.1× SSC, 0.5% SDS at room temperature for 10 min, followed by incubation in 50% formamide, 0.5% SDS, 0.01× SSC at 47 °C for 1 hour. After 3 washes with 2× SSC, 0.5% SDS at room temperature for 5 min (to remove formamide), the blot should be clean. For a harsher cleaning procedure the blot may be washed several times in 0.01× SSC, 0.5% SDS at 90–95 °C. However, very strong signals will "survive" even this drastic procedure. Thus, when multiple hybridizations of the same blot with different probes are required, we recommend using those probes which produce the strongest signals at the end.

Blots may be stored in 2× SSC, 0.1% SDS, 1 mM NaN_3, or in prehybridization solution in a sealed plastic bag at 4 °C.

Southern blot

Several procedures are available for DNA blotting and subsequent hybridization with biotinylated probes (3, 11), and all seem to yield results of comparable quality. Thus, we present only 1 short protocol which has been successfully used in our lab for detection of single-copy genes in yeasts and higher plants.

Genomic DNA is digested with appropriate restriction enzymes and loaded onto a TAE-agarose gel (0.7–1.0%, 4 mm thickness). After electrophoresis and photodocumentation of the ethidium bromide-stained gel, the destained gel is depurinated for 6 min in 0.25 M HCl, and then denatured twice for 15 min in 0.5 M NaOH, 1 M NaCl, and twice for 10 min in 5× SSC, 10 mM NaOH. Capillary blotting is performed with 5× SSC, 10 mM NaOH (transfer buffer) for 30–40 min. The assembly for capillary transfer is the same as described above for Northern blots, except that the weight on top of the paper towels should be only 1 g/cm^2.

Following transfer the procedures for UV cross-linking, hybridization, and blot development are essentially identical to those described for Northern blots. The incubation in conjugation buffer can be reduced to 20 min, and the exposure to X-ray film usually needs not more than 30–60 min.

The same blot can be used 3–5 times before background is unacceptable. Stripping is done by 1 wash in 0.1× SSC, 0.5% SDS for 10 min at room temperature, followed by an incubation for 30–60 min at 37 °C in 0.2 M NaOH, 1 M NaCl, and twice in 2× SSC, 0.1% SDS at room temperature for 10 min for neutralization. Storage of the blots is essentially as described for Northerns.

Quantitative analysis by densitometry

Quantitative analysis of signals obtained with chemiluminescent substrates encounters a number of technical problems not met when radiolabeled probes are used. While in the latter case liquid scintillation counting allows direct signal quantitation, the use of nonradioactive detection systems requires densitometric signal analysis on X-ray film. If only relative signal intensities are to be compared, the area units obtained with a densitometric scanner must be proportional to the amount of target. This is usually achieved by serial dilutions of samples (dot blot or slot blot) until signal intensity is in the linear range (12). As an alternative, exposure time can be varied. Note that since nonradioactive detection is based on enzymatic activity of alkaline phosphatase, small temperature differences during blot development may affect signal intensity considerably. However, a quantitative comparison of samples on separately developed blots is possible when standard samples are included to allow normalization between different blots. Alternatively, a standard blot with a known serial dilution of target sequence may be co-developed in the same solutions under identical conditions.

While the procedures described above yield quantitative comparisons, they do not allow assessment of absolute amounts of bound probe, since label incorporation during probe synthesis is random. However, when probes are synthesized by PCR with one of the primers biotinylated at its 5' end, even an absolute quantitation of bound probe molecules is possible.

12.3 Results and Discussion

Figure 12.1 shows a Southern blot of genomic DNA of the plant *Daucus carota*. Genomic DNA was isolated according to ref. 13. The DNA was CsCl purified (7) and digested with 10 units/µg DNA with the indicated restriction enzymes. The digests were carried out at 37 °C for 3–6 hours, with a DNA concentration of 50 µg/ml. The same blot was sequentially hybridized with 3 different probes for the detection of V-type H^+-ATPase genes. The results indicate the presence of 1 or at maximum 2 isoforms for subunits A and B, while the quite complex restriction pattern for the c subunit points toward 3 to 4 isoforms.

Probe sizes were 700 bp (A subunit), 280 bp (c subunit), and 160 bp (B subunit), respectively. As can be seen from the comparison, the stripping procedure efficiently removed all signals produced during the previous hybridization. Also, reprobing had no effect on signal strength (exposure time was identical after each blot development). Background became slightly darker (compare signals for A and B subunits to signals for the c subunit) but remained tolerable.

Figure 12.2 shows a Northern blot of total RNA samples isolated (9) from different organs of *Daucus carota* plants exposed to salt stress. The blot was hy-

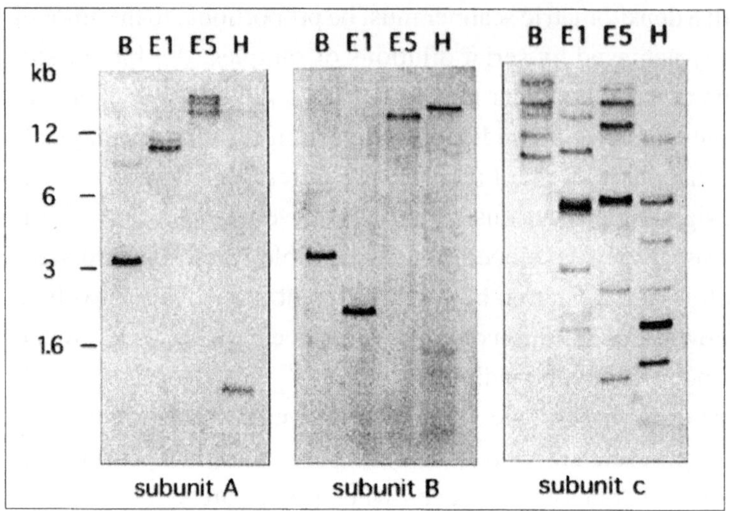

***Figure 12.1 Southern-blot analysis of V-type H^+-ATPase genes of* Daucus carota**

Three micrograms of genomic DNA was loaded per lane. Hybridization was in 30% formamide, 1% SDS, 1 M NaCl, 6% PEG 6000, 250 µg salmon sperm DNA, and 0.1 nM probe. Stringency conditions during hybridization and high-stringency wash were based on ≥65% sequence homology. The order of hybridization was (i) c-subunit, (ii) A-subunit, (iii) B-subunit probe. Exposure time to X-ray film (Kodak X-OMAT) was 30 min. Restriction enzymes were Bam HI (B), EcoRI (E1), Eco RV (E5), and Hind III (H)

bridized with probes for the A subunit and the c subunit of the V-type H^+-ATPase. The two hybridizations were performed sequentially, using optimized conditions for each hybridization. After the first hybridization with the 700-bp probe for the A subunit, the blot was subjected to a high-stringency wash (0.2× SSPE, 0.5% SDS, at 60–62 °C for 60 min), then washed twice in 2× SSPE, 0.5% SDS, and again prehybridized (60 min at 42 °C) and hybridized overnight with the second probe (c subunit, 280 bp). The second high-stringency wash was done in 0.2 × SSPE, 0.5% SDS, at 58 °C. Using this protocol, no unspecific signals were observed.

An 18S-RNA probe was used to confirm equal sample loading. Here, no further stripping is necessary prior to hybridization, as the exposure time for 18S-RNA detection is only 5–10 min, compared with about 2 hours for detecting mRNAs for subunits A and c.

The results shown are part of a study on the expression of different V-type H^+-ATPase subunits in salt-stressed plants. The V-type H^+-ATPase is thought to play a

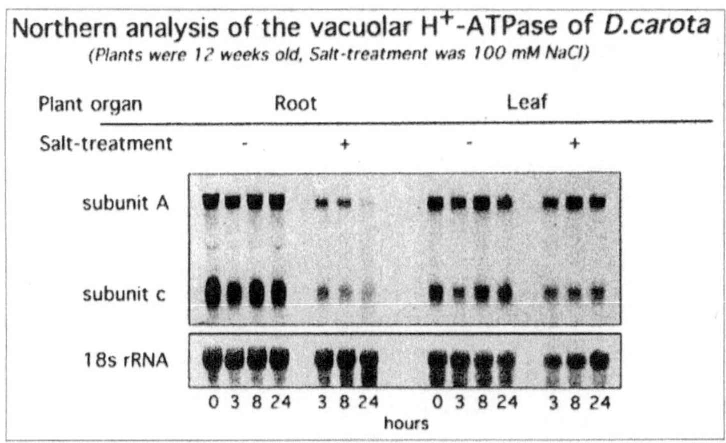

Figure 12.2 Northern-blot analysis of V-type H+-ATPase mRNA levels (subunits A and c) in roots and leaves of Daucus carota plants exposed to 100 mM NaCl for 3, 8, and 24 hours

Ten micrograms of total RNA was loaded per lane. Hybridization was done in 50% formamide, 5% SDS, 1 M NaCl, 6% PEG 6000, 250 µg/ml of salmon sperm DNA, and 0.1 nM probe at 42 °C. The stringency during hybridization was based on 75% sequence homology, while the high-stringency washes were based on 90% homology. Exposure time to X-ray film (Hyperfilm ECL, Amersham) was 2 hours for mRNAs of subunits A and c, and 5 min for 18S RNA.

role in vacuolar salt sequestration. Note the dramatic decrease of transcript levels for the subunits A and c in the root of this salt-sensitive crop. In the leaf the transcript level of subunit A remains unaffected, whereas the transcript level of subunit c is slightly decreased (14).

The densitometric analysis of nonradioactive blots (Fig. 12.3) shows that special care has to be taken to assure a linear correlation between bound probe and signal strength. For demonstration we have blotted a 5'-labeled PCR fragment of known quantity to a nylon membrane and determined the area units of the X-ray signals obtained after different exposure times. Analysis was done with a Quadra 800 computer (Macintosh) using the densitometry program NIH 1.57. The results show that with increasing exposure time the useful linear range shifts toward lower concentrations. The data confirm that for a certain exposure time, which has to be determined empirically, a linear correlation between target and signal intensity (area units) is obtained.

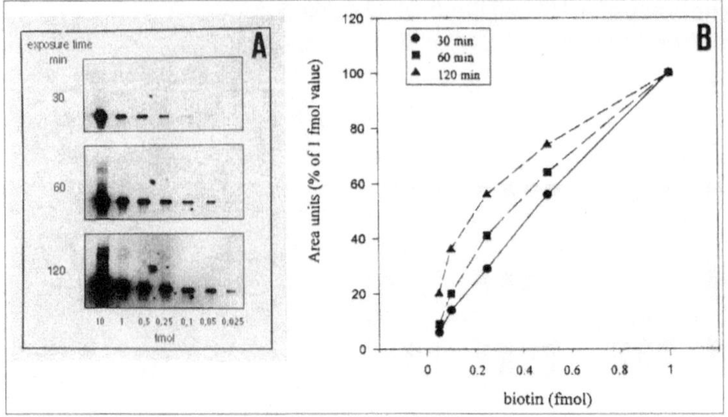

Figure 12.3 Quantitative evaluation of a 5'-labeled PCR fragment blotted to a nylon membrane

(A) signals obtained by X-ray film exposure; (B) densitometric determination of signals obtained after different film exposure times. Area units for 1 fmol were 2236, 4148, and 7613 after exposure times of 30, 60, and 120 min, respectively. Signals for 1 fmol were arbitrarily set to 100%.

12.4 Troubleshooting

No or weak signal

Overestimation of loaded DNA/RNA: load higher amount of DNA/RNA.
Incomplete transfer: If the amount of remaining nucleic acids in the gel is too high, a prolonged transfer time may help. If gel compression during transfer is too rapid, the weight should be reduced. On the other hand, if transfer is too slow, the weight may be increased. As a rule of thumb, if using a sponge, after 1 hour of transfer 1–2 cm of the previously dry paper towels should be soaked with transfer buffer.
Low incorporation of labeled nucleotide into the probe: Insufficient labeling is not necessarily a matter of low concentration of biotin-labeled nucleotide in the mixture but of the capability of the enzyme to incorporate it (for more information see ref. 15 and literature cited therein). Therefore different ratios of dTTP/bio-dUTP (other than 1:1) or equivalent nucleotides usually solve these problems.

Background problems

Background is spotty: This is usually caused by aggregated AP conjugate. A 1-min centrifugation of the AP conjugate at 12 000 g prior to use is helpful.
Cloudy, irregular surface with lots of spots: (i) The blocking reagent (namely, casein) may be too old, causing bad blocking, or (ii) the membrane may have been contaminated with bacteria prior to UV cross-linking (if this is the case, the blot is no longer useful). To avoid these problems, containers should be as clean as possible (try SDS and NaOH washes). (iii) The hybridization solution is contaminated, or the salmon sperm DNA is used up. To solve this problem, set up a new hybridization solution.

References

1 Lanzillo, J.J. (1991) *Anal. Biochem.* **194**, 45–53.
2 Klevan, L. and Gebeyehu, G. (1990) *Methods Enzymol.* **184**, 561–577.
3 Bronstein, I., Voyta, J.C., Lazzari, K.G., Murphy, O., Edwards, B. and Kricka, L.J. (1990). *BioTechniques* **8**, 310–314.
4 Löw, R. and Rausch, T. (1994) *BioTechniques* **17**, 1026–1030.
5 Lo, D.Y-M., Mehal, W.Z. and Fleming, K.A. (1990) *In*: Innis, M.A., Gelfand, D.H., Sninsky, J.J., White, T.J. (eds.) *PCR Protocols: A Guide to Methods and Applications*. Academic Press, pp 113–118.
6 Emanuel, J.R. (1991) *Nucl. Acid Res.* **19**, 2790.
7 Ausubel, F., Brent, R., Kingston, R.E., Moore, D.D., Seidman, J.G., Smith, J.A. and Struhl, K. (1987–1995) *Current Protocols in Molecular Biology*. Vol. 1, Greene Publishing.
8 Sambrook, J., Fritsch, E.F. and Maniatis, T. (1989) *Molecular Cloning: A Laboratory Manual*. Second Edition, Cold Spring Harbor Laboratory Press.
9 Logemann, J., Schell, J. and Willmitzer, L. (1987) *Anal. Biochem.* **163**, 16–20.
10 Meinkoth, J. and Wahl, G. (1984) *Anal. Biochem.* **138**, 267–284.
11 Leary, J.J., Brigati, D.J. and Ward, D.C. (1983) *Proc. Natl. Acad. Sci. USA* **80**, 4045–4049.
12 Walsh, P.S., Varlaro, J. and Reynolds, R. (1992) *Nucl. Acid Res.* **20**, 5061–5065.
13 Murray, M.G. and Thompson, W.F. (1980) *Nucl. Acid Res.* **8**, 4321–4325.
14 Löw, R., Rockel, B., Kirsch, M., Ratajczak, R., Lüttge, U. and Rausch, T. (1994) *In*: *Abstracts of the 4. Intern. Congress of Plant Mol. Biol., Inter. Soc. of Plant Mol. Biol.*, 404.
15 Helmuth, R. (1990) Nonisotopic Detection of PCR-Products. *In: PCR Protocols: A Guide to Methods and Applications. Academic Press*, pp 119–128.

13 Biotin-Labeled Riboprobes to Study RNA-Binding Proteins

Pedro L. Rodriguez and
Luis Carrasco

Summary

Most of the methods available to detect and study RNA-binding proteins use a radioactively labeled component. We describe a nonradioactive modification of the Northwestern assay using biotinylated riboprobes. The method detects binding of biotinylated RNA to proteins fixed on nitrocellulose. Proteins are first separated by SDS-polyacrylamide gel electrophoresis (PAGE), transferred to the nitrocellulose membrane, and renatured. Therefore, generation of variant proteins and the analysis of their RNA-binding capacity can be quickly accomplished. The biotinylated riboprobes are stable and easy to handle, permitting the processing of large numbers of samples in a short time. Chemiluminescence is generated with streptavidin-conjugated peroxidase and luminol-luciferin. This method has been successfully applied to the study of poliovirus RNA-binding proteins.

13.1 Introduction

Proteins endowed with RNA-binding activity play important functions in the metabolism of nucleic acids and the regulation of gene expression (1, 2). This diverse group of proteins comprises at least 9 families which have been distinguished on the basis of their RNA recognition motifs (3). The identification of new members of RNA-binding proteins and the characterization of their RNA-binding requirements usually involves extensive mutagenesis analysis, followed by protein purification and assays of their interactions with nucleic acids. Hence, a rapid method for protein purification, as well as a rapid and reliable test for RNA binding, would facilitate these types of studies.

Several techniques have been used to evaluate the interaction of proteins with RNA. Most of these methods use a radioactively labeled component. Assays that use radioactively labeled RNA include nitrocellulose filter binding (4), gel retardation (5), cross-linking by UV light (6), and Northwestern blotting (7). This last method detects protein-RNA interactions by binding labeled RNA to proteins fixed to nitrocellulose (8). Northwestern analysis is widely used to examine the RNA-binding capacity of proteins (9–11), but exhibits the drawbacks of any procedure requiring the use of radioactivity. Apart from methods based on radioactively labeled RNA, assays that employ radioactively labeled proteins have been described (12, 13). To our knowledge, the first report of an RNA-binding assay that uses biotinylated RNA was described by Scherly et al. (12), who tested binding of [^{35}S]methionine-labeled protein to biotinylated RNA. Binding of human U1-specific A protein to *Xenopus* snRNA transcripts with biotinylated UTP incorporated was detected by precipitation of the biotinylated RNA-protein complex with streptavidin-agarose beads (12). The RNA-bound protein was eluted from the nucleic acid by boiling, separated on an SDS-PAGE gel, and visualized by fluorography. A variation of this streptavidin-biotin-mediated precipitation assay has been described by Ashley et al. (14) using magnetic beads (instead of agarose beads) with conjugated streptavidin to capture the biotinylated RNA-protein complex. Using this assay, RNA-binding activity was demonstrated with protein FMR1 involved in the fragile X syndrome (14). To test the specificity of RNA binding, *in vitro* transcription of a human cDNA library in the presence of biotin-UTP was performed. Thus, a complex mixture of biotinylated RNAs was produced, some of which bound to FMR1 (14).

We reasoned that the combination of a classic Northwestern assay with the capacity of a protein to bind biotinylated RNA could be exploited to detect proteins that bind RNA, avoiding the use of radioactive reagents. Thus, a nonradioactive modification of the Northwestern assay using biotinylated riboprobes is described. The application of this technique to the analysis of 2 poliovirus proteins involved in genome replication, proteins 2C and 3AB, has determined that both proteins bind RNA (15, 16). The analysis of the RNA-binding capacity of a number of variant proteins has elucidated the amino acid residues of proteins 2C and 3AB involved in RNA binding (15, 16).

13.2 Technical Procedures

Materials

Enzymes and pMal-c vector were purchased from New England Biolabs (Beverly, MA, USA). Streptavidin-conjugated peroxidase, luminol, luciferin, and nucleotides were from Boehringer Mannheim (Mannheim, Germany). Biotin 21-UTP was purchased from Clontech (Palo Alto, CA, USA). Other reagents used were from Sigma (St. Louis, MO, USA).

Preparation of biotinylated riboprobes

As general references for synthesis of RNA probes by *in vitro* transcription of double-stranded DNA templates, the reader is referred elsewhere (17–19). Bacteriophage DNA-dependent RNA polymerases can incorporate biotin-labeled NTPs into RNA (20), opening the possibility of replacing radioactively labeled riboprobes with nonhazardous biotinylated riboprobes. The protocol described by Clontech laboratories can be used to prepare biotin-labeled riboprobes, where biotin-labeled NTP is incorporated into RNA by an *in vitro* transcription, with the modification that the transcription mixture contains a 1:1 ratio of biotin 21-UTP to UTP. Biotin 21-UTP is a UTP analog which has a biotin attached to the pyrimidine ring by a 21-atom spacer arm, which minimizes steric hindrance (Clontech). Other reports make use of biotin 11-UTP that is incorporated into RNA by *in vitro* transcription using a 1:10 ratio of biotin 11-UTP to UTP (12, 14).

Protocol 13.1

Synthesis of biotinylated riboprobes

1. Add the following reagents to an RNase-free Eppendorf tube placed on ice in the following order: DEPC-treated H_2O, 50 µl final volume; 10 µl of 5× transcription buffer (final volume: 40 mM Tris-HCl (pH 7.5); 6 mM $MgCl_2$; 2 mM spermidine; 10 mM NaCl); 5 µl of 100 mM DTT; 1 µml of RNasin inhibitor (Promega); 5 µl of 10 mM ATP; 5 µl of 10 mM CTP; 5 µl of 10 mM GTP; 2.5 µl of 10 mM UTP; 2.5 µl of 10 mM biotin 21-UTP; 1 mg of linearized riboprobe template DNA (Promega or Stratagene vector); 50 units of bacteriophage RNA polymerase (SP6, T3 or T7 RNA polymerase, depending on the vector and orientation chosen). In our hands SP6 yields less RNA than T3, or T7 RNA polymerases. To avoid transcription of undesirable sequences, linearize the template DNA with a restriction enzyme that leaves 5′ overhangs or blunt ends. After digestion with the restriction enzyme, purify the DNA by phenol/chloroform extraction and subsequent ethanol precipitation.
2. Mix gently and incubate at 37 °C for 90 min.
3. To remove the DNA template, add RQ1 RNase-free DNase (Promega) to a concentration of 1 unit/µg DNA. Incubate at 37 °C for 10 min.
4. Remove nonincorporated biotinylated ribonucleotide by Sephadex G-50 gel filtration. Recovery of biotinylated riboprobe is nearly quantitative. Do not use phenol extraction. Biotinylated probes are soluble in the phenol layer.
5. Agarose gel electrophoresis and ethidium bromide staining under RNase-free conditions provides an estimate of the concentration of RNA obtained by comparison of the biotinylated riboprobe to the DNA markers. The amount of newly synthesized biotinylated riboprobe depends on the amount, size (site of linearization), and purity of the template DNA; the nucleotide concentration does not become limiting during the *in vitro* transcription. Under standard conditions, at least 5 mg of biotinylated RNA is transcribed from 1 mg of template DNA.
6. Store the biotinylated probe at –20 °C. It is stable for at least 1 year.

Protocol 13.2

Northwestern assay

1. SDS-PAGE and blotting. Separate the proteins by SDS-PAGE under standard conditions (21). Do not boil the proteins before loading them. One lane containing molecular mass standards (Bio-Rad) should be included. The proteins are transferred overnight from the gel to a nitrocellulose membrane using a wet electrotransfer protocol as described previously (21).
2. Check the efficiency of protein transfer by Ponceau-S staining (21). Mark the position of the proteins on the filter with a pen. This mark does not affect the renaturation of the proteins, and the Ponceau-S staining will disappear in the following steps.
3. For renaturation of the proteins wash the filter with buffer A (10 mM Tris-HCl (pH 7.5), 1 mM EDTA, 50 mM NaCl, 0.1% Triton X-100, 0.02% bovine serum albumin (BSA), 0.02% Ficoll 400, and 0.02% PVP) 3 times, 30 min each, with rocking at room temperature in the same buffer. All subsequent incubations and washes are performed using the same solution. The blotted proteins are then probed with biotinylated riboprobe at 40 ng/ml for 1 hour. Unbound RNA is removed by 3 cycles of washing, 2 min per wash.
4. Detection of RNA-binding proteins. Incubate the membrane for 30 min with streptavidin-conjugated peroxidase (Boehringer) (1:20 000) and then wash the filter 3 times for 2 min with buffer A. Finally, visualize the RNA binding proteins by placing the filter in the H_2O_2-luminol-luciferin solution (Sigma) (22), composed of 0.1 M Tris-HCl (pH 8.0), 1.25 mM luminol, 30 μM luciferin, 2 mM H_2O_2, for 1 min, air-dry the membrane on Whatman 3MM paper, and seal it in a polyethylene bag. The resulting light signal is recorded on Kodak film by contact exposures (usually 30–300 sec).

The various steps of the procedure are summarized in Figure 13.1.

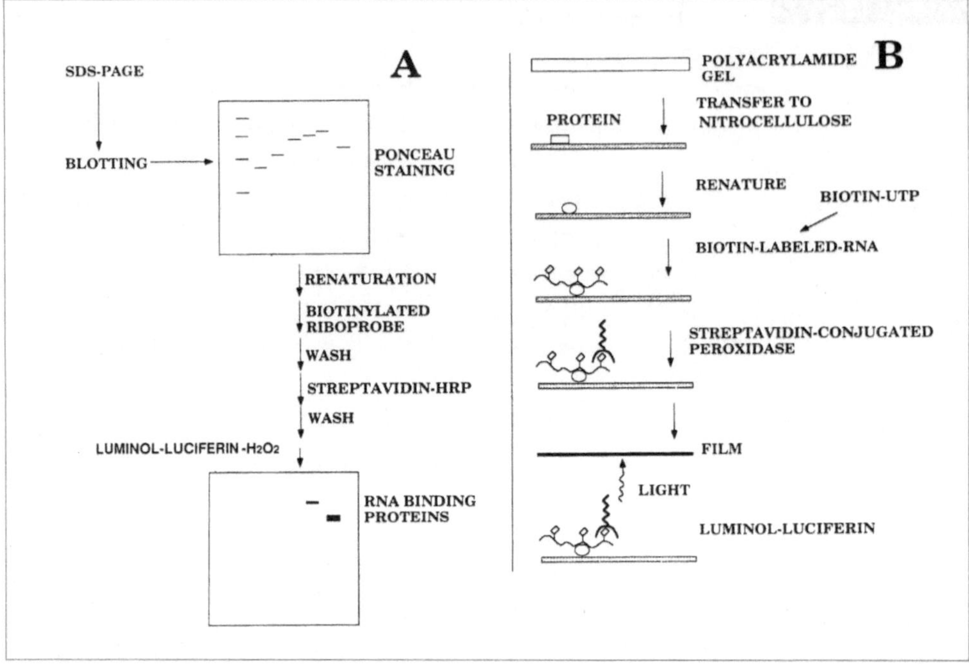

Figure 13.1 Schematic diagrams of the Northwestern assay procedure

13.3 Results and Discussion

Poliovirus protein 2C expressed as a fusion protein with the maltose-binding protein (MBP) gives rise to a retardation complex in polyacrylamide gels after association with a partially double-stranded RNA substrate, while MBP alone is devoid of this activity (23). A deletion mutant MBP-2C (1-255) lacking the C-terminal region between amino acids 256 and 329 loses the capacity to interact with RNA (23). Further analysis of the RNA-binding properties of 2C has been carried out using the nonradioactive Northwestern assay described above. As a first step to validate the technique, a comparison of the radioactive v. nonradioactive Northwestern analysis was carried out (Fig. 13.2). CI is an RNA-binding protein from plum-pox virus with RNA helicase activity (24). Poliovirus protein 2C has NTPase activity and also binds RNA (23). MBP and MBP-Dbgal were chosen as control proteins that, in principle, lack RNA-binding capabilities. Figure 13.2 shows that biotinylated RNA was efficiently bound by MBP-2C and CI proteins transferred to nitrocellulose. The result was quantitatively similar to that observed with the radioactive probe, although higher sensitivity is observed with the nonradioactive assay. Thus, more intense bands of bound RNA are apparent in the MBP-2C and CI. Moreover, a truncated product of CI observed by Ponceau-S staining has RNA-binding capacity as detected by biotinylated RNA. The specificity of the assay is observed in Figure 13.2, panel C, since both MBP and MBP-Dbgal failed to bind RNA. Moreover, 2 deletion mutants of poliovirus protein 2C, MBP-2C (1-255) and MBP-2C (1-297), are devoid of RNA-binding capacity.

The results obtained agree well with the RNA-binding properties of the MBP-2C and CI proteins observed using gel-retardation assays (23, 25). Moreover, they confirm that a portion of the 2C protein beyond amino acid 297 is involved in RNA binding (23). It appears, therefore, that the last 32 amino acids of the carboxy terminus control the interaction of 2C with RNA. A smaller deletion of 10 amino acids (Fig. 13.3) has no influence on the RNA-binding capacity of 2C, thereby locating the RNA-binding activity to residues 298–319. There is a sequence in this region

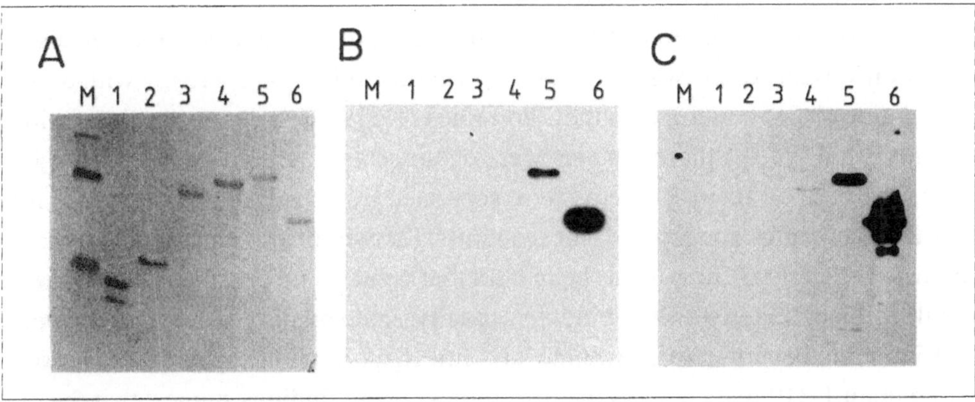

Figure 13.2

(A) Ponceau-S staining

Lane 1, MBP (maltose-binding protein); lane 2, MBP-Dbgal; lane 3, MBP-2C (1-255); lane 4, MBP-2C (1-297); lane 5, MBP-2C; lane 6, CI. M is the molecular mass standards. The purification of MBP fusion proteins was carried out by affinity chromatography on amylose resin columns as described (30). The proteins were subjected to SDS-PAGE, blotting, and Ponceau-S staining as described under Technical Procedures.

(B) Radioactive Northwestern assay

A radioactive riboprobe encompassing nt 2099–4600 poliovirus RNA was used as the probe. The riboprobe was synthesized from a Bluescript KS (Stratagene, San Diego, CA, USA) plasmid containing a 2500-bp fragment of the poliovirus cDNA. Lanes 1–6 as indicated in (A).

(C) Nonradioactive Northwestern assay

A biotinylated riboprobe encompassing nt 2099–4600 poliovirus RNA was used as the probe. Lanes 1–6 as indicated in (A).

(Taken with permission from (30))

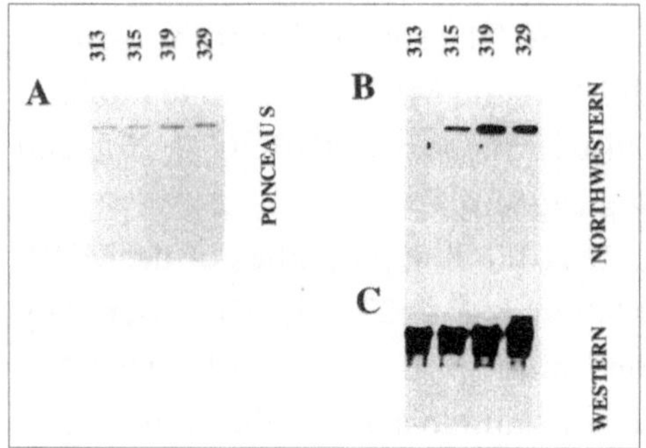

Figure 13.3
(A) Ponceau-S staining
313, 315, 319, and 329 are described in the text and correspond to MBP_2-2C fusion proteins with the indicated amino acids of 2C.
(B) Northwestern assay *of the fusion proteins indicated in (A). A biotinylated riboprobe encompassing nt 2099–4600 poliovirus RNA was used as the probe.*
(C) Western immunoblot assay *carried out with anti-2C antiserum. The same nitrocellulose sheet containing the fusion proteins analyzed in the Northwestern assay was subsequently analyzed in the Western immunoblot assay.*

that is rich in arginines and, in this respect, resembles the RNA-binding domains described for other proteins which are endowed with this activity (6, 26–29). We decided, therefore, to mutate the sequence NERNRR (amino acids 312–317) in order to determine its importance for RNA binding. After PCR-based mutagenesis, 3 deletion variants of 2C were obtained, 1 without arginines (1-313), 1 with a single arginine (1-315), and 1 with the 3 arginines (1-319). The corresponding fusion proteins (MBP-2C) of these mutants were obtained and purified as indicated (Fig. 13.2). The purified fusion proteins were separated by SDS-PAGE and assayed for their RNA-binding capacity and, subsequently, for their reactivity with anti-2C antibodies. Figure 13.3 shows that the mutant that contains no arginines in the C-terminal region is devoid of RNA-binding capacity, some binding is detected with the 1-315 mutant that contains 1 arginine, and full RNA-binding capacity is observed with mutant 1-319 that contains the 3 arginines. These findings suggest that the region between residues 313 and 319 of 2C is essential for RNA binding. The motif NERNRR of 2C is similar in sequence to those described in other proteins with RNA-binding properties (6, 27–29). Moreover, analogous sequences, rich in basic amino acids, not only are located in this region of poliovirus 2C but are found in other 2C proteins from other picornaviruses (23), suggesting, again, an involvement of such sequences in RNA-binding activity. A complete analysis of the RNA-binding activity of poliovirus protein 2C using biotinylated riboprobes in a Northwestern assay has recently been described (15).

The nonradioactive Northwestern assay has also been used to examine the RNA-binding activity of poliovirus protein 3AB and different 3AB variants obtained by a PCR-based random mutagenesis method (16). The overexpression of these proteins in *Escherichia coli* yields large amounts of each of the variant proteins. Crude extracts of these *E. coli* clones, when analyzed by SDS-PAGE, show a major band of the overexpressed protein that is clearly distinguishable from the other proteins in the gel. Therefore, the Northwestern assay allows a preliminary analysis of mutated 3AB proteins in total lysates without further purification (16). These analyses have permitted identification of the amino acid residues of protein 3AB involved in RNA binding (16).

13.4 Troubleshooting

A general troubleshooting guide concerning Western blotting protocols and chemiluminiscent detection is provided by Amersham accompanying the ECL system for Western blotting detection and nucleic acid analysis. Additional advice to avoid problems specifically associated with the use of biotinylated riboprobes and/or the Northwestern assay is as follows.

Biotinylated riboprobes

1. Remember that artefactual initiation of transcription has been observed from 3' overhanging termini and, to a lesser extent, from blunt ends (18). Therefore, linearize the DNA template preferably by using a restriction enzyme that leaves 5' overhangs. After the restriction digest, purify the DNA by phenol/chloroform extraction and subsequent ethanol precipitation.
2. Purification of the biotinylated riboprobe can be accomplished by chromatography on Sephadex G-50 columns, or by appropriate spin columns or ethanol precipitation, but do not use phenol extraction, as biotinylated probes are soluble in the phenol layer.
3. To check the efficiency of biotinylation, prepare dilutions of biotinylated riboprobe at concentrations of 1000, 100, 10, and 0 pg/µl, and spot 1 µl of each dilution directly onto a strip of nitrocellulose filter. Fix the RNA onto the membrane by UV irradiation, and follow the Northwestern assay from the addition of the biotinylated riboprobe to the detection of the chemiluminescent signal.

Northwestern assay

1. SDS-PAGE. Separate the proteins by SDS-PAGE under standard conditions. The electrophoresis is performed at a current of 30–40 mA, depending on the characteristics of the acrylamide gel. Higher currents are not advised, since we have found that the RNA-binding capability of the protein is destroyed if high-voltage electrophoresis is carried out.
2. Transfer of the proteins from the gel to the nitrocellulose filter is performed wet, because partially dry transfer causes heating of the samples and may inhibit subsequent renaturation of the transferred proteins (7).
3. The proteins must be allowed to renature for at least 90 min by washing the filter in buffer A.
4. If the positive control does not bind RNA, check the previous steps. For best results, freshly prepared samples should be used. Make sure that all buffers are RNase-free.
5. If nonspecific background signal is obtained, reduce the riboprobe concentration or increase the duration of the washing steps after riboprobe incubation and streptavidin-conjugated peroxidase incubation. Make sure that the blot or anything that contacts the blot is clean and fingerprint-free.

Acknowledgments

The expert technical assistance of Mr M.A. Sanz is acknowledged. Protein CI was a generous gift of Dr J.A. Garcia. Plan Nacional project number BIO 92-0715, DG-ICYT project number PB90-0177, and the institutional grant to the CBM of Fundación Ramón Areces are acknowledged for their financial support.

References

1 Kenan, D.J., Query, C.C. and Keene, J.D. (1991) *TIBS* **16**, 214–220.

2 Biamonti, G. and Riva, S. (1994) *FEBS Lett.* **340**, 1–8.

3 Mattaj, I.W. (1993) *Cell* **73**, 837–840.

4 Méthot, N., Pause, A., Hershey, J.W.B. and Sonenberg, N. (1994) *Mol. Cell. Biol.* **14**, 2307–2316.

5 Osman, T.A.M., Thömmes, P. and Buck, K.W. (1993) *J. Gen. Virol.* **74**, 2453–2457.

6 Pause, A., Méthot, N. and Sonenberg, N. (1993) *Mol. Cell. Biol.* **13**, 6789–6798.

7 Chen, X., Sadlock, J. and Schon, E.A. (1993) *Biochem. Biophys. Res. Commun.* **191**, 18–25.

8 Bowen, B., Steinberg, J., Laemmli, U.K. and Weintraub, H. (1980) *Nucl. Acids Res.* **8**, 1–20.

9 Wakefield, L. and Brownlee, G.G. (1989) Nucl. Acids Res. **17**, 8569–8580

10 McCormack, S.J., Thomis, D.C. and Samuel, C.E. (1992) *Virology* **188**, 47–56.

11 Boyle, J.F. and Holmes, K.V. (1986) *J. Virol.* **58**, 561–568.

12 Scherly, D., Boelens, W., van Venrooij, W.J., Dathan, N.A., Hamm, J. and Mattaj, I.W. (1989) *EMBO J.* **8**, 4163–4170.

13 Boelens, W.C., Jansen, E.J.R., van Venrooij, W.J., Stripecke, R., Mattaj, I.W. and Samuel, I.G. (1993) *Cell* **72**, 881–892.

14 Ashley, C.T., Jr., Wilkinson, K.D., Reines, D. and Warren, S.T. (1993) *Science* **262**, 563–566.

15 Rodriguez, P. and Carrasco, L. (1995) *J. Biol. Chem.* **270**, 10105–10111.

16 Lama, J., Sanz, M.A. and Rodríguez, P.L. (1995) *J. Biol. Chem.* **270**, 14430–14438

17 Krieg, P.A. and Melton, D.A. (1984) *Nucl. Acids Res.* **12**, 7057–7070.

18 Krieg, P.A. and Milton, D.A. (1987) *Methods Enzymol.* **25**, 397–415.

19 Sambrook, J., Fritsch, E.R. and Maniatis, T. (1989) *Molecular Cloning: A Laboratory Manual.* Cold Spring Harbor Laboratory Press, NY.

20 Luehrsen, K.R. and Baum, M.P. (1987) *BioTechniques* **5**, 660–670.

21 Harlow, E. and Lane, D. (1988) *Antibodies: A Laboratory Manual.* Cold Spring Harbor Laboratory, NY.

22 Whitehead, T.P., Thorpe, G.H.G., Carter, T.J.N., Groucutt, C. and Kricka, L.J. (1983) *Nature* **305**, 158–159.

23 Rodríguez, P.L. and Carrasco, L. (1993) *J. Biol. Chem.* **268**, 8105–8110.

24 Lain, S., Riechmann, J.L. and García, J.A. (1991) *Nucl. Acids Res.* **18**, 7003–7006.

25 Lain, S., Martín, M., Riechmann, J.L. and García, J.A. (1991) *J. Virol.* **65**, 1–6.

26 Lazinski, D., Grzadzielska, E. and Das, A. (1989) *Cell* **59**, 207–218.

27 Calnan, B.J., Tidor, B., Biancalana, S., Hudson, D. and Frankel, A.D. (1991) *Science* **252**, 1167–1171.

28 Kiledjian, M. and Dreyfuss, G. (1992) *EMBO J.* **11**, 2655–2664.

29 Lee, C.-Z., Lin, J.-H., Chao, M., McKnight, K. and Lai, M.M.C. (1993) *J. Virol.* **67**, 2221–2227.

30 Rodríguez, P.L. and Carrasco, L. (1994) *BioTechniques* **17**, 702–705.

Rapid YAC End Sequencing by Alu-Vector PCR and Biotinylated Primers

Ricardo Fujita and Anand Swaroop

Summary

Yeast artificial chromosome (YAC) clones are widely used to isolate genomic DNA (ranging from 100 to 2000 kilobase pairs) from the region of interest. Terminal sequences from YACs are isolated to verify if the clone is nonchimeric and to generate sequence-tagged sites for mapping and isolation of overlapping clones. These end sequences can be selectively amplified by Alu-vector polymerase chain reaction (PCR); this method uses a primer corresponding to the consensus of Alu sequences interspersed in the human genome and another derived from the YAC vector. We have modified the Alu-vector PCR procedure by altering the vector primer sequence and by biotinylation of this primer. Two thermocycle profiles are performed: the annealing temperature during the first 10 cycles is 60 °C, which favors polymerization from the vector; this is followed by 25 cycles with 45 °C annealing temperature to facilitate the synthesis from any Alu sequence present in the proximity of the vector. The PCR-amplified product is then incubated with streptavidin-conjugated magnetic beads, allowing the rapid purification of specific Alu-vector sequences from other contaminating fragments (e.g., those obtained by inter-Alu PCR and/or from the host *Saccharomyces cerevisiae*). Two strands of the amplified Alu-vector PCR product are then separated and sequenced without further purification.

14.1 Introduction

Identification of genes responsible for inherited diseases by the "positional cloning" strategy has been facilitated dramatically because of the development of molecular tools to isolate large genomic DNA regions and improved methods for the selection of transcribed sequences. A genetic disease locus is first mapped to a particular chromosome or a chromosomal region by linkage analysis or cytogenetic rearrangements in affected families. Subsequently, the genomic region (of several hundreds of kb) spanning the disease locus is isolated in yeast artificial chromosomes (YACs), which are the most adequate and preferred vehicles for cloning DNA of this size (1). YAC technology has facilitated the task of mapping, cloning, and analyzing expression of complex genomes and is widely used in positional cloning of disease genes. Huntington's disease (2) and spinal muscular atrophy (3) genes are recent examples of gene isolation with YACs. In general, a collection of contiguous YAC clones is used to cover the critical region of a disease gene. Although randomly isolated markers from a YAC can be tested on other clones, the use of markers from both YAC ends is more efficient in searching for contiguous clones (4). Approaches to isolate YAC ends include subcloning by conventional methods or by end rescue (1) and polymerase chain reaction (PCR) amplification after ligation to linkers (5) or self-ligation for inverse PCR (6). These techniques require several steps before the purification of YAC-end DNA.

A simpler protocol, called Alu-vector PCR, is based on the presence of Alu repeat sequences interspersed at an average of 4 kb in the human genome (7). In this method, one primer corresponds to the Alu consensus sequence, and the other is anchored in the YAC vector (pYAC4, constructed with yeast and pBR322 DNA; the insert is in the Sup4 gene of *S. cerevisiae)*. Nonspecific products are common artifacts in Alu-vector PCR, and include Alu-Alu products and those from the Sup4 gene. As in other methods, specific end sequences need to be isolated by analytical and preparative gel electrophoresis.

We have modified the Alu-vector PCR by (i) altering the vector primer sequences from Sup4 to the pBR322 regions of the vector, therefore eliminating the risk of Sup4 amplification, and (ii) biotinylation of the vector primer for quick, specific, and efficient isolation of amplified products using solid-phase magnetic capture (8). Using this approach, we were able to purify and sequence the ends of YAC clones from the Xp21.1 chromosomal region (9).

14.2 Technical Procedures

Isolation of YAC DNA

The method for obtaining yeast (or YAC) DNA generally involves the preparation of protoplasts (see ref. 10 for an improved miniprep protocol). However, we routinely isolate YAC DNA from agarose blocks that are used for pulsed-field gel electrophoresis (PFGE).

Materials and buffers

AHC medium (Protocol 14.1)

- 1.7 g yeast nitrogen base w/o amino acids and ammonium sulfate, dehydrated (DIFCO, Detroit, MI)
- 5 g ammonium sulfate (Sigma, St. Louis, MO)
- 10 g caseine acid hydrolysate (acid) (Hy case amino) (ICN Biochemicals, Cleveland, OH)
- 20 mg adenine hemisulfate salt (Sigma, St. Louis, MO)
- Add water up to 950 ml, adjust to pH 5.8 with HCl and autoclave
- Add 50 ml of filtered 40% glucose (w/v)

Protocol 14.1

Isolation of YAC DNA

1. Inoculate YAC clone into 5 ml of AHC medium and incubate at 30 °C in a shaker at 250 rpm for 24–30 hours until the culture becomes light pink (only transformed yeasts produce pink cultures).
2. Centrifuge the culture tube for 15 min at 3500 g. Discard the supernatant.
3. Prepare a 2% low-melting-point agarose (GIBCO BRL, Gaithersburg, MD) gel in 50 mM EDTA and keep it at 42 °C.
4. Wash cells twice with 10 ml of 50 mM EDTA and resuspend pellet in an equal volume of 50 mM EDTA. Incubate at 37 °C for 5 min.
5. Mix equal volumes of resuspended cells and liquid agarose thoroughly and cast immediately into chilled PFGE block

molds or a chilled petri dish. Let it harden for 20 min at 4 °C. At this stage, the agarose blocks can be cut to fit into 15-ml tubes.

6. Resuspend the blocks in 10 volumes of solution containing 0.5 M NaCl, 0.25 M EDTA, 0.125 M Tris-HCl (pH 7.5), 0.5 M β-mercaptoethanol, 1% sarkosyl, and proteinase K at 2 mg/ml. Incubate at 50 °C for 2 hours to overnight.
7. Incubate blocks twice with phenylmethylsulfonyl fluoride (PMSF) (Sigma, St. Louis, MO, USA) at 40 µg/ml in 50 mM EDTA for 30 min at 42 °C to inactivate proteinase K.
8. Rinse blocks twice with 50 mM EDTA (at this point blocks can be stored at 4 °C or used for PFGE; for DNA preparation, proceed to step 9).
9. Rinse agarose blocks in 10 volumes of TE (10 mM Tris-HCl (pH 8.0), 1 mM EDTA). Equilibrate with 10 volumes of 1x agarase buffer (New England Biolabs, Beverly, CA) in a microfuge tube on ice for 30 min.
10. Discard the buffer and melt blocks at 65 °C for 5 min and then incubate at 37 °C for 5 min.
11. Add 1 unit of β-agarase (New England Biolabs, Beverly, CA), mix thoroughly by tapping the tube, spin for 2 sec in microfuge, and incubate for 1 hour to overnight at 37 °C.
12. Chill microfuge tube on ice for 15 min, and pellet the remaining solid particles by centrifugation at 13 000 rpm for 5 min.
13. Transfer supernatant to a fresh tube and load 10 µl in an agarose gel to determine the quantity of high molecular weight DNA (higher than 20 kb in an 0.7% agarose gel). Dilute a fraction of the DNA to 2 ng/µl for PCR amplification.

Alu-vector PCR

Oligonucleotide primers

Biotinylated and nonbiotinylated primers were synthesized at the DNA synthesis Core of the University of Michigan Medical School. Vector primers are designed from the pYAC4 vector (1): from the pBR322 sequence, and from the Sup4 region. A schematic representation of the YAC left arm is shown in Figure 14.1; primers

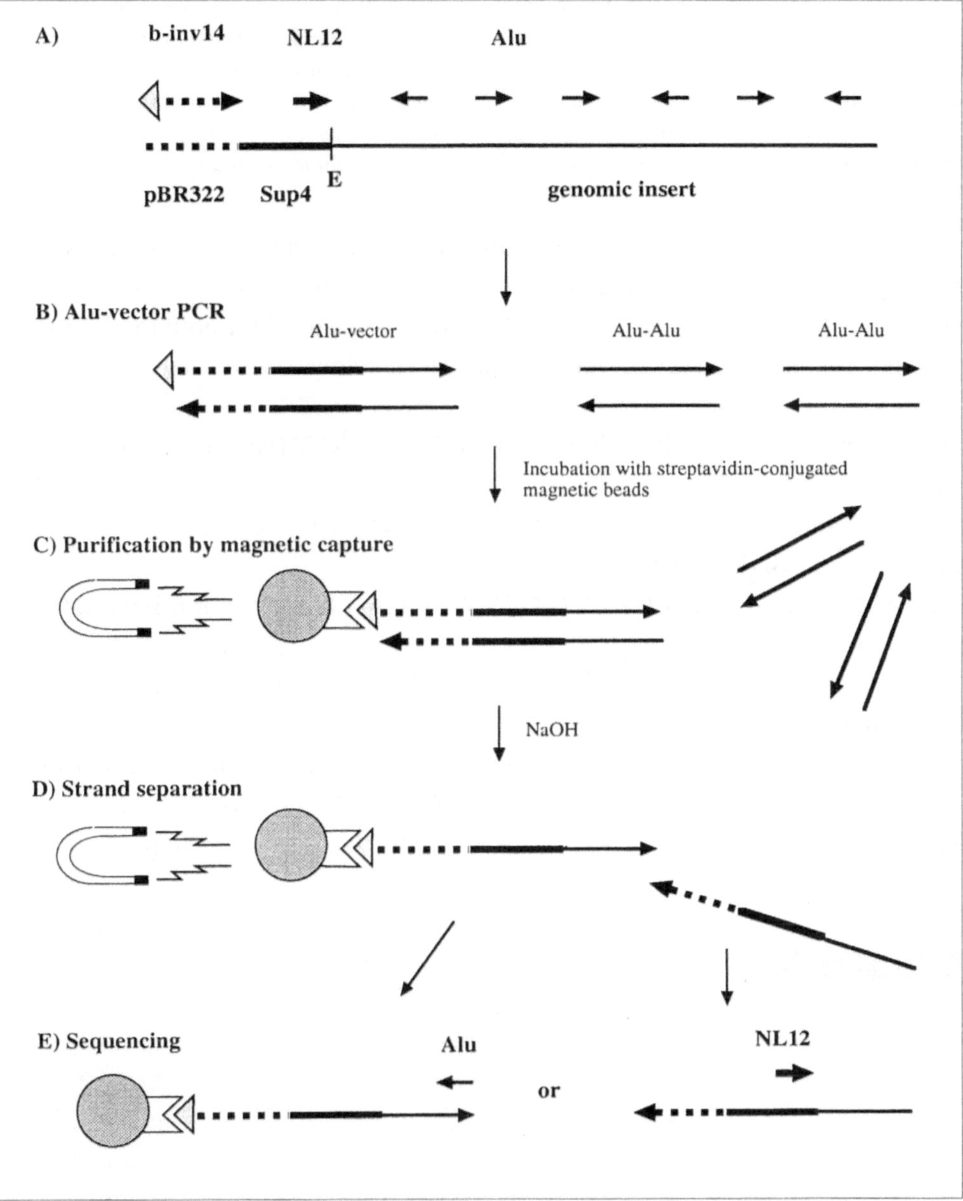

Figure 14.1 Schematic representation of the modified Alu-vector PCR method for isolating YAC end sequences

(A) *Representation of vector YAC left arm and its boundaries. Dashed, thick and thin lines, and arrows depict pBR322, Sup4, and insert segments, respectively. Lines indicate YAC DNA and arrows indicate primers, b-inv14 is a biotinylated primer. E is the EcoRI insert site.*

(B) *Alu-vector PCR is performed with biotinylated b-inv14 and an Alu primer. Segments carrying vector sequences (on the left) will amplify as well as*

To Figure 14.1 (continued). Alu-Alu products (on the right). ***(C)*** *Biotinylated product is incubated with streptavidin-conjugated magnetic beads and separated from Alu-Alu products by magnetic capture.* ***(D)*** *Purified YAC-end double-strand DNA is denatured, the biotinylated strand remains attached to magnetic beads, and the nonbiotinylated strand is in solution. (E) Sequencing is performed separately for each strand: Alu primer is used to sequence the biotinylated strand, and nested vector primer NL12 (from Sup4) is used for the nonbiotinylated strand.*

from the right arm are in similar position but opposite orientation. The pBR322-derived primers are biotinylated and used for PCR amplification, whereas nested Sup4 primers are used for sequencing. Alu primers are used for both PCR and sequencing. Sequences of various primers are given below:

b-inv14: 5'biotin- ATG CGC ACC CGT TCT CGG AGC 3' (vector primer, left; 314 bp from the insert)
b-inv20: 5'biotin- ATG CCG GCC ACG ATG CGT CCG GCG 3' (vector primer, right; 121 bp from the insert)
NL12: 5' CAA TTA AAT ACT CTC GGT AGC CAA G 3' (vector primer, left; 31 bp from the insert)
NL10: 5' CTC CCG GGG GCG AGT CGA ACG CCC 3' (vector primer, right; 20 bp from the insert)
NA71: 5' CCT CGG CCT CCC AAA GTG CTG GGA TTA CAG 3' (Alu primer)
NH34: 5' AAG TCG CGG CCG CTT GCA GTG AGC CGA GAT 3' (Alu primer)
NG76: 5' CGA CCT CGA GAT CTC GGC TCA CTG CAA 3' (Alu primer)

Protocol 14.2 **Alu-vector PCR reaction**

1. Add in an ice-chilled 0.5-ml microfuge tube: 40.6 µl of water, 2 ng of YAC DNA (1 µl), 5 µl of 10× PCR buffer (100 mM Tris-HCl, pH 8.2, 500 mM KCl), 1 µl of 50 µM biotinylated vector primer, 1 µl of 20 µM Alu primer, 1 µl of 10 mM of each dNTP, and 0.4 µl of AmpliTaq 5 units/µl (Perkin-Elmer, Norwalk, CT).
2. Use another tube with all the components except vector primer as control (Alu-Alu reaction).
3. Cover the reaction mixture with 50 µl of mineral oil and put in a thermocycler machine. Use the following programs in sequence for amplification:

Program	°C	Time	Cycles
#1	94	5 min	1
#2	94	45 sec	10
	60	45 sec	
	72	2 min	
#3	94	45 sec	25
	45	45 sec	
	72	2 min	
#4	72	7 min	1

4. Load 10 µl of the PCR reaction in a 1% agarose minigel and electrophorese until the bromophenol blue dye reaches two-thirds of the gel (about 40 min in the TAE buffer using 90 volts). Multiple bands should be evident in the Alu-vector reaction. These products should be different from those obtained in the control Alu-Alu reaction. Sometimes a unique segment may be amplified in the Alu-vector reaction with no product obtained with the Alu primer alone.
5. Proceed to the solid-phase purification step.

Protocol 14.3

Solid phase purification

1. Transfer 20 µl of Dynabeads to a l.5-ml microfuge tube and place it in a magnetic tray (Dynal MPC-E, Dynal, Oslo) for 30 sec. Remove the supernatant with a pipette while keeping the tube in the tray.
2. Wash Dynabeads twice: resuspend particles in 40 µl of B&W solution (10 mM Tris-HCl, pH 7.5, 1 mM EDTA, 2 M NaCl), place the tube back in the tray for 30 sec to separate the beads from the solution, and remove the supernatant with the tube in the tray.
3. Resuspend Dynabeads in 40 µl of B&W solution and mix with 40 µl of Alu-vector PCR reaction.
4. Incubate mixture with gentle agitation at 37 °C for 30 min to allow streptavidin-biotin interaction to take place.
5. Wash twice with B&W solution. At this stage, vector-specific DNA is attached to Dynabeads. Nonspecific PCR products remain in solution and are removed during the washes.
6. Resuspend Dynabeads in 8 µl of 0.1 N NaOH and incubate at

room temperature for 10 min. Captured Alu-vector-specific DNA is denaturated and separated. At this stage, the biotinylated-strand of the Alu-vector PCR product remains attached to Dynabeads, and the nonbiotinylated strand is in solution.

7. Replace tube on the magnetic tray, and collect the supernatant into a fresh tube kept on ice. Label tubes with biotinylated (Dynabeads) and nonbiotinylated (supernatant) DNA strands.
8. Wash Dynabeads once with 50 µl of 0.1 N NaOH, twice with 50 µl of TE, and finally resuspend them in 20 µl of water for sequencing.
9. Neutralize the supernatant from step 6 by adding 4 µl of 0.2N HCl and 1 µl of 1 M Tris-HCl, pH 7.5.
10. Sequence 7 µl of the single-stranded DNA with 0.5 pmol of the oligonucleotide primer using a Sequenase kit (US Biochemicals, Cleveland, OH). Alu primer is used for sequencing the biotinylated strand, and a nested vector primer (derived from the Sup4 sequence) for the complementary nonbiotinylated strand. Three µl of the sequencing reaction is loaded on a 6% polyacrylamide gel containing 50% urea (Sequagel, National Diagnostics, Atlanta, GA).

Biotinylated amplified products are captured using streptavidin-conjugated magnetic beads (Dynabeads M280, Dynal, Oslo), as recommended by the manufacturer for direct solid-phase sequencing.

14.3 Results and Discussion

We have applied this procedure to sequence the ends from YAC clones isolated from the genomic region spanning the X-linked retinitis pigmentosa (RP3) and ornithine transcarbamylase (OTC) genes at Xp21.1. Figure 14.2 shows an example of typical products obtained after amplification using the YAC DNA with Alu primer either alone (control) or with vector primers. The comparison of 2 sets of PCR products shows that many bands are unique in each reaction, although some common products are observed between the Alu only and the Alu-vector PCR reactions. The presence of nonspecific Alu products does not interfere with the isolation of Alu-

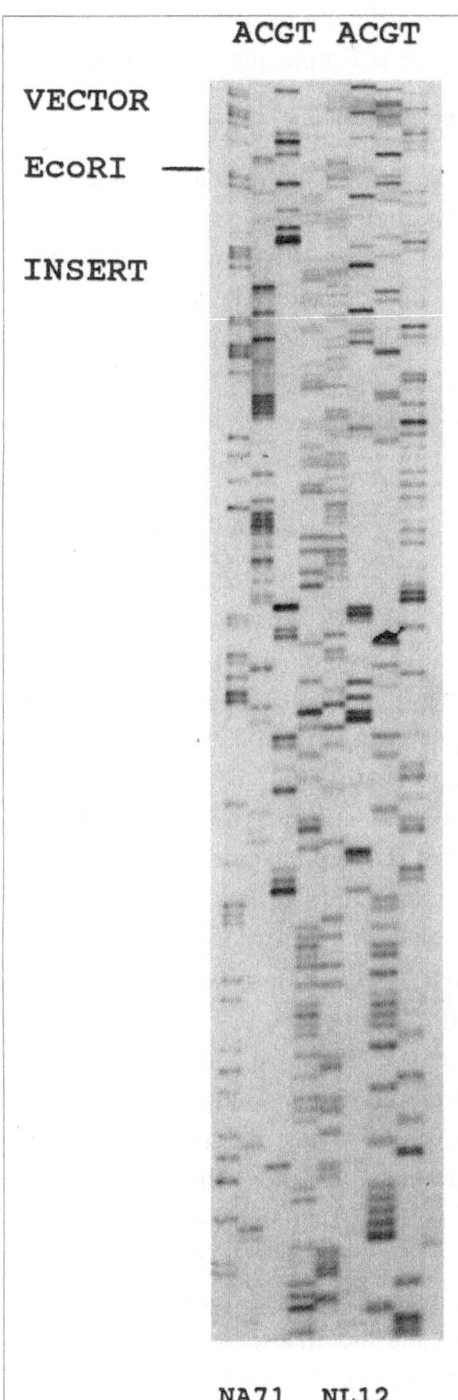

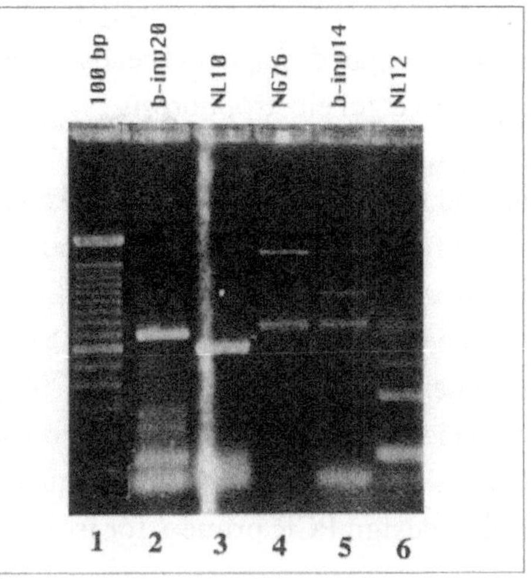

▲
Figure 14.2 Fragments produced by Alu-vector PCR

YAC OTC-C DNA was amplified with the Alu primer NG76 and the vector primers, as indicated in Technical Procedures. Lane 1, 100 base-pair ladder (GIBCO BRL); lane 2, NG76 Alu primer + biotinylated vector primer b-inv20; lane 3, NG76 + nested vector primer NL10; lane 4, NG76 primer alone; lane 5, NG76 + biotinylated vector primer b-inv14; and lane 6, NG76 + nested vector primer NL12. Differences between the amplification with Alu primer alone and those with the vector primers are clearly observed; however, some common bands (probably Alu-Alu products) are also seen (e.g., the 700-bp band is present in both lanes 4 and 5).

◀ **Figure 14.3 Sequence of the selected Alu-vector products**

YAC OTC-C DNA was amplified with the Alu primer NA71 and the biotinylated vector primer, b-inv14, corresponding to the YAC left arm. Sequence on the left was obtained from the biotinylated strand using the Alu primer NA71. The Eco RI insert site as well as the regions corresponding to insert and vector are indicated. Sequence obtained from the nonbiotinylated strand with nested vector primer NL12 is shown on the right. Sequences from the 2 strands reveal antiparallel complementarity.

vector PCR products, since only these will be specifically retained by the streptavidin-conjugated magnetic beads. This procedure is simple and does not involve preparative gel electrophoresis.

Figure 14.3 shows the sequence of 2 complementary strands obtained from a purified Alu-vector PCR product. The 2 sequences reveal antiparallel complementarity of these strands and show the expected boundaries between the vector and the insert. These results were confirmed by determining the sequence of cloned YAC end fragments. The captured YAC end products can also be cloned and used as probes for mapping purposes. It should be possible to extend this method to other species by using primers derived from species-specific interspersed repeat sequences. The genomic YAC end sequences obtained in these experiments have been used to design PCR primers for isolating overlapping clones from YAC libraries.

14.4 Troubleshooting

We use one biotinylated vector primer for 3 reactions, each containing a different Alu primer to enhance the probability of appropiate amplification. Agarose gel electrophoresis of PCR products reveals if the amplification has been successful and which primer pair is better. Failure to obtain YAC end sequence (and any PCR product) is due primarily to the poor quality of the YAC DNA preparation. Also, in some instances it may be necessary to optimize the appropriate DNA amount or $MgCl_2$ concentration for PCR amplification (e.g., use 0.2–20 ng YAC DNA or 0.5–4 mM $MgCl_2$ for PCR reaction, and check the products by gel electrophoresis). PCR amplification can also be optimized using Alu primers alone.

If there are specific fragments after Alu-vector PCR but the purified DNA does not work for sequencing reactions, verify the biotinylation of amplified products. After gel electrophoresis, amplified products can be transferred to a nylon membrane by Southern blotting and biotinylated products identified using streptavidin-based systems (e.g., Phototope Detection Kit, New England Biolabs, Beverly, MA).

Because of a random distribution of Alu repeats, YACs lacking Alu sequences in close proximity of the vector arms may not yield specific Alu-vector PCR products (11). Primers derived from other interspersed repeat sequences (e.g., Kpn repeats) or short random sequence primers may be used to overcome this problem.

Longer Alu-vector products can be obtained using additives such as glycerol, formamide, Taq Extender (Stratagene, La Jolla, CA) in PCR reactions. New long-range PCR methods should also facilitate the amplification of larger Alu-vector PCR products (12).

Acknowledgments

We thank Dr A. Hanauer (Strasbourg, France) for the b-inv20, b-inv14, and NH34 primer sequences. This research was supported by a postdoctoral fellowship from Fight for Sight (Schaumburg, IL) to R.F., and by grants from the National Institutes of Health (EY07961) and the Retinitis Pigmentosa Foundation Fighting Blindness, Baltimore, to A.S.

References

1 Burke, D.T., Carle, G.F. and Olson, M.V. (1987) *Science* **236**, 806–812.

2 The Hungtinton's Disease Collaborative Research Group (1993) *Cell* **72**, 971–983.

3 Lefebvre, S., Bürglen, L., Reboullet, S., Clermont, O., Burlet, P., Viollet, L., Benichou, B., Cruaud, C., Millasseau, P., Zeviani, M., Le Paslier, D., Frezal, J., Cohen, D., Weissenbach, J., Munnich, A. and Melki, J. (1995) *Cell* **80**, 155–165.

4 Palazzolo, M.J., Sawyer, S.A., Martin, C.H., Smoller, D.A.I and Hartl, D.L. (1991) *Proc. Natl. Acad. Sci. USA* **88**, 8034–8038.

5 Riley J., Butler, R., Ogilvie, D., Finniear, R., Jenner, D., Powell, S., Anand, R., Smith, J.C. and Markham A.F. (1990) *Nucl. Acids Res.* **18**, 2887–2890.

6 Silverman, G.A., Ye, R.D., Pollock, J.E., Sadler, J.E. and Korsmeyer, S. (1989) *Proc. Natl. Acad. Sci. USA* **86**, 7485–7489.

7 Nelson, D.L., Ledbetter, S.A., Corbo, L.A., Victoria, M.F., Ramirez-Solis, R., Webster, T.D., Ledbetter, D.H. and Caskey, C.T. (1989) *Proc. Natl. Acad. Sci. USA* **88**, 6157–6161.

8 Hultman, T.S., Bergh; S., Moks, T. and Uhlen, M. (1991) *BioTechniques* **10**, 84–93.

9 Fujita, R. and Swaroop, A. (1995) *BioTechniques* **18**, 796–799.

10 Lee, F.-J.S. (1992) *BioTechniques* **12**, 677.

11 Silverman, G.A. (1993) *PCR Methods and Applications* **3**, 141–150.

12 Barnes, W.M. (1994) *Proc. Natl. Acad. Sci. USA* **91**, 2216–2220.

Index

Acryl-azido group 138
Adhesion molecule 100
Affinity chromatography 41
- avidin-streptavidin 65
Alu repeat 228
Alu-vector PCR 228
Antibody 41, 89, 90
Avidin 2
Avidin-agarose 55
Avidin-binding protein 47, 52

Biotin, photoactivable 25
Biotin, succinimide ester of (BNHS) 173
Biotin hydrazide 13, 118
Biotin labeling 92, 131
Biotin-binding sites of SA 173
Biotin-BMCC 17
Biotin-HPDP 17, 48
Biotin-LC-hydrazide 13
Biotin-lysine-tRNA^{Lys} 183, 185, 188–191, 193
Biotin-7-NHS 85, 86, 87, 93, 94
Biotinamidocaproate-*N*-hydroxy-succinimide ester 118
5-(Biotinamido)pentylamine 16
Biotinylated
- antibodies 151, 154, 167
- aprotinin 156
- ligand 48
- membrane protein 91
- PCTI-1 151, 154, 156
- poly(ADP-ribose) 131
- SLPI 151, 154, 155
- trypsin 151, 154, 158

Biotinylation 65, 85 - 87, 89, 92, 93, 100
- of aprotinin 146
- of aprotinin-trypsin complex 146
- of PCTI-1 146
- of protein 84
- of RBC 174
- of serine proteinases 144
- of SPI 144
- of SLPI 146
- of trypsin 147
- of trypsin-SBTI complex 147
- , On-dish 91
Biotycin-hydracide 13
Blot development 206
Blot reprobing 207

Cell membrane preparation 87
Cell surface proteins 83, 84
Chemiluminescence 83, 85, 89–91, 95
Cholecystokinin 38
Complement 167, 168

Densitometry 208
Digitonin 47, 51
Dithio-bis(succinimidylpropionate)(DSP) 100
Dithio-bis(sulfo-succinimidylpropionate)(DTSSP) 100
Dog pancreas microsomes 183, 193–195
Drug targeting 167, 168

EGF-receptors (EGF-R) 88, 90, 91
Electrophoretic mobility shift assay 193, 195
Ellmann's reagent 7
Enhanced chemiluminescence (ECL) 65, 100

Epidermal growth factor receptor (EGF-R) 83, 85
ExtrAvidin 117

Genomic DNA 228
Glycophorin A 117
Glycoprotein 116

HABA 4
Histones 139
Horseradish peroxidase (HRP) 95
Hybridization 205

Immunoerythrocytes 167, 168, 177
Immunoprecipitation 83–85, 88, 90–92
Insulin 65
Iodination, of protein 83
Iodoacetyl-LC-biotin 17

Lectin 115
Lectinoblotting 121
Ligand blotting 131
Ligand-receptor complex, covalent 65
Liver uptake 178
Lymphocyte 99
Lysolecithin 100

Mapping 228
Membranes
- Immobilon-PVDF 95
- Nitrocellulose 89, 95
- Nylon 95
Membrane fraction 85, 87, 90–92
Membrane protein 87
Microtiter plate assay 115, 119
Monoclonal antibody 85

Neutravidin 2
NHS-biotin 10
NHS-iminobiotin 10
NHS-LC-biotin II 10
NHS-SS-biotin 9
o-Nitrobenzyl ester 38
Nonradioactive detection 99
Nonradioactive Northwestern assay 220
Northern blot 202
Northwestern blotting 216

Oligosaccharide 116
Oligosaccharyl transferase 193

PCR-labeling of (c)DNA probes 202
Permeabilization 99
Photoaffinity labeling 67
Photobiotin 132
Photoelution 40
Pituitary adenylate cyclase-activating polypeptide (PACAP) receptor 46
Poliovirus protein 2C 220
Poly(ADP-ribose) binding protein 131
Poly(ADP-ribose) polymerase 133
Poly(ADP-ribose) protein interaction 131
Poly(ADP-ribose) preparation 131
Ponceau S 95
"Positional cloning" 228
Preabsorption 94
Protease inhibitors 93
Protein A-agarose 94
Protein G-agarose 88, 89, 94
Protein, nuclear 138

Rabbit reticulocyte lysate 184, 196
Radioiodination 71
Radiolabeling of protein 83, 84
Receptor binding-site 65
Receptor purification 46
Receptor-binding assay 57
Red blood cells (RBC) 167, 168
Reticulocyte lysate 183, 185, 190, 192
RNA-binding 217
RNA-binding domain 222
RNA-binding protein 215
RNA-recognition motif 216

SATA 18
Serum 176

Signal peptidase 193
Solid-phase magnetic capture 228
Southern blot 208
SPDP 18
Streptavidin 2, 89, 167, 169
Streptavidin peroxidase 89
Streptavidin peroxidase conjugate 83, 91
Sulfo-NHS-biotin 9, 94
Sulfo-NHS-LC-biotin 9

Taq polymerase 202
Transcription, *in vitro* 185, 186, 190
Transcription factors 193
Translation, *in vitro* 92, 183, 184, 187, 191 196
Translation kit, *in vitro* 195
Traut's reagent 18

Western blot 83
Western blotting 115, 185, 187, 194, 196
Western transfer 85, 89, 91, 95

YAC end 228
Yeast artificial chromosomes (YACs) 228